Microrheology with Optical Tweezers

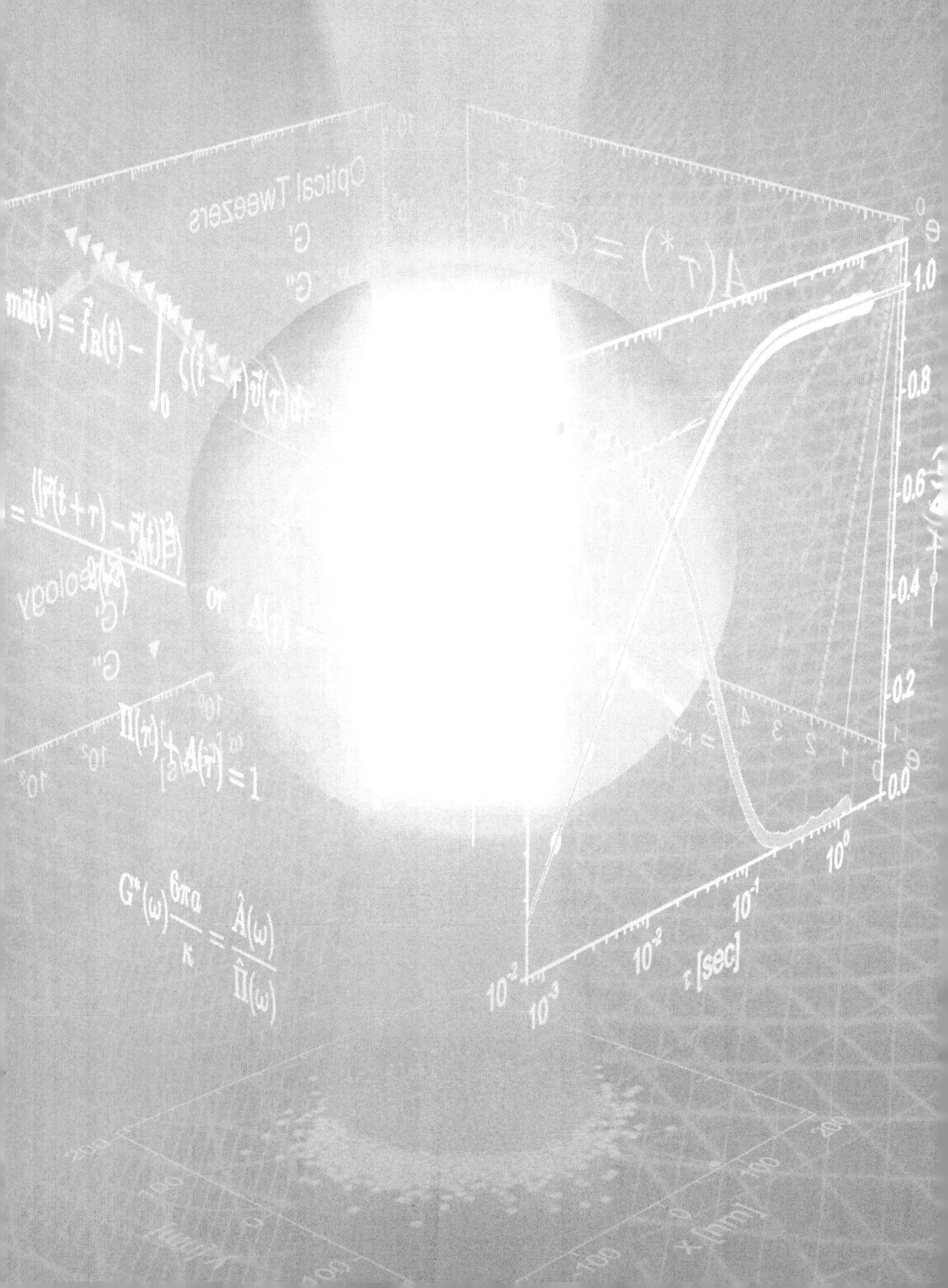

Microrheology with Optical Tweezers

Principles and Applications

edited by

Manlio Tassieri

Pan Stanford Publishing

Published by

Pan Stanford Publishing Pte. Ltd.
Penthouse Level, Suntec Tower 3
8 Temasek Boulevard
Singapore 038988

Email: editorial@panstanford.com
Web: www.panstanford.com

British Library Cataloguing-in-Publication Data
A catalogue record for this book is available from the British Library.

Microrheology with Optical Tweezers: Principles and Applications

ISBN 978-981-4669-18-4 (Hardcover)
ISBN 978-981-4669-19-1 (eBook)

Printed in the USA

To Loredana

Contents

Part IV Review on Optical Tweezers Applications

Foreword

Following more than a decade of related pioneering research, Arthur Ashkin and co-workers (Ashkin et al., 1986) in 1986 published the seminal paper that inspired the field of optical tweezers. Optical tweezers use tightly focussed beams of laser light to trap transparent, micrometer-sized particles, typically suspended in a fluid. That these micrometer-sized particles could be trapped and manipulated using only a single laser beam meant that optical tweezers could easily be implemented as a simple addition to a conventional microscope. This simplicity of construction makes optical tweezers an accessible tool for further research. The ability to use optical tweezers for measurements of picoNewton forces and nanometer displacements is of particular interest to the biophysics community. For example, by using optical tweezers, Block and co-workers were able study the properties of individual motor proteins (Svoboda et al., 1993).

The physical sciences too have embraced optical tweezers and in 1995 Rubinsztein-Dunlop and co-workers (He et al., 1995) demonstrated that by using specially modified laser beams they could transfer angular momentum between the light and the particle, causing the particle to spin. This conversion of an optical tweezer to an optical spanner both answered fundamental questions about the nature of light and suggested new approaches to the creation of optically driven micromachines.

A major technical advance in optical tweezers was championed by David Grier and co-workers (Grier, 2003). They used a single laser but added a spatial light modulator to give the user real-time control of both the structure of the trapping beam and indeed the number of independent traps. This simple addition of one extra optical component creates holographic optical tweezers capable of simultaneously trapping many particles at defined positions within a sample volume.

The use of optical tweezers to measure both forces and position relies on their ability to measure the position of the centre of mass of the trapped probe-particle to a small fraction of the optical wavelength. Initially these measurements used quadrant photodiodes, aligned to the axis of the particle, to measure the slight deflection in the transmission of the trapping laser beam. More recently similar performance has been achieved using high-speed video cameras, which are both easier to align and capable of tracking multiple particles simultaneously.

Being able to measure the higher frequency vibrations of the trapped particles means that the particle motion can be related to the viscoelastic properties of the fluid in which the particle(s) is (are) suspended. These measures of local viscosity, microrheology, have been applied within microfluidic channels and inter- or even intracell environments and the tracking of multiple particles reveals the eigenmodes of particle networks and the precise nature of their hydrodynamic coupling and potential synchronisation.

As cameras, spatial light modulators, and computing technologies advance, the complexity of the systems that can be both controlled and probed will also advance. These advances will cover systems which encompass the synchronisation of motor proteins, the folding of complex molecules, and the creation of digital colloids.

Professor Miles J. Padgett
Kelvin Chair of Natural Philosophy
and Vice Principal (Research)
University of Glasgow
Glasgow, UK

Editor's Preface

The aim of this book is to provide a pedagogical introduction to the physics principles governing optical tweezers and their application in the field of microrheology.

This is achieved by following a linear path that starts from a narrative introduction of the "nature of light", followed by a rigorous description of the fundamental equations governing the propagation of light through matter. In particular, the ability of light to exert a force on micrometer-sized objects is described, this property being the fulcrum of the optical tweezers' working principles. Moreover, some of the many possible instrumental configurations are presented, especially those that better adapt to perform microrheology measurements.

In order to better appreciate the microrheological methods with optical tweezers explored in this book, informative introductions to the basic concepts of linear rheology, statistical mechanics, and the most popular microrheology techniques are also given.

Furthermore, an enlightening prologue to the general applications of optical tweezers different from rheological purposes is provided at the end of the book.

I am grateful to all the co-authors for their generous contributions to this book. I have a debt of gratitude to both Professor Jonathan M. Cooper and Professor Miles J. Padgett for their continuous guidance and support during the editing of this book. I would like to thank Niall P. MacDonald for the book cover. The book has been edited/written during the tenure of a Royal Academy of Engineering/The Engineering and Physical Sciences Research Council Research Fellowship. Finally, I wish to thank the University of Glasgow for the supportive and enthusiastic work environment that has facilitated the writing of this book.

Dr Manlio Tassieri
RAEng/EPSRC Research Fellow
Division of Biomedical Engineering
School of Engineering
University of Glasgow
Glasgow, UK

Part I

Introduction

Chapter 1

General Introduction to Optical Tweezers and Their Applications

Manlio Tassieri
Division of Biomedical Engineering, School of Engineering, University of Glasgow, Glasgow G12 8LT, UK
manlio.tassieri@glasgow.ac.uk

1.1 Introduction

It is thanks to the pioneering works of Ashkin and co-workers (Ashkin, 1970; Ashkin and Dziedzic, 1971; Ashkin et al., 1986) that optical tweezers (OT) have become an invaluable tool for a myriad of studies throughout the *natural sciences*. Indeed, the ability of manipulating micron-sized objects by means of a highly-focussed laser beam have proved to be an extraordinary tool for a great number of applications, as reviewed, for instance, in (Curtis et al., 2002; Grier, 2003; Moffitt et al., 2008; Molloy and Padgett, 2002; Novotny et al., 1997; Svoboda and Block, 1994a).

The main reason of their success is based on the fact that they can be considered as exceptionally sensitive transducers able to resolve pN forces and nm displacements, with high temporal

Microrheology with Optical Tweezers: Principles and Applications
Edited by Manlio Tassieri

ISBN 978-981-4669-18-4 (Hardcover), 978-981-4669-19-1 (eBook)
www.panstanford.com

resolution, down to μs. This has encouraged a multitude of scientists to adopt optical tweezers as extraordinarily sensitive tools for measuring phenomena occurring at *time, length* and *force scales* traditionally unaccessible by conventional methods. Notably, it is a fortuitous coincidence that a very wide range of biological processes fall within the OT range of sensitivity. Indeed, biophysicists have been the firsts to take advantage of such exceptional tools; e.g., OT have been successfully employed for measuring the compliance of bacterial tails (Block et al., 1989), the forces exerted by a single motor protein (Finer et al., 1994), the mechanical properties of human red blood cells (Yoon et al., 2008) and those of individual biological molecules (Ashkin and Dziedzic, 1987; Mehta et al., 1999; Tskhovrebova et al., 1997).

The continuous development of technology has significantly expanded the potential applications of OT. In particular, the development of special optical devices like the spatial light modulators (SLM) has contributed to the expansion of *holographic optical tweezers* apparatus and applications (e.g., Curtis et al., 2002; Dufresne and Grier, 1998; Dufresne et al., 2001; Fournier et al., 1995; Hayasaki et al., 1999; Liesener et al., 2000; Reicherter et al., 1999)), which in conjunction with recent advances in camera technology (Gibson et al., 2008; Keen et al., 2007), has augmented the OT ability of trapping multiple particles simultaneously and to track their positions at frame rates of *kHz* for indefinite periods of time (Di Leonardo et al., 2007b). The ability of trapping, positioning and tracking many particles at the same time has been found of particular use in colloidal science (Crocker, 1997; Crocker and Grier, 1994; Curran et al., 2012; Dinsmore et al., 1996; Meiners and Quake, 1999), where the hydrodynamic interactions either between particles *or* with the surrounding environment are of great importance in understanding colloid phenomena such as sedimentation (Lee et al., 2014) and aggregation (Curran et al., 2014).

Another great advantage of adopting fast multi-particle position detectors in OT setups is given by the ability of studying hydrodynamic interactions occurring between particles trapped in fluids with very low viscosity, such as air (Di Leonardo et al., 2007a; Yao et al., 2009b), allowing investigation into aerosols. But also, they have allowed the observation of a parametrically excited resonance within a Brownian oscillator (Di Leonardo et al., 2007a) and underdamped modes in periodic arrays of trapped particles (Yao et al., 2009b).

The inherent scale of optical tweezers makes them readily applicable to microfluidic systems. These systems offer many advantages over conventional methods for biological and chemical measurements. The miniaturisation of micro-analytical devices results not only in a low fabrication cost and a reduction in the volume of potentially expensive reagents used, but also in an increased speed of analysis and the ability to run multiple analytical processes in parallel. In recent years there have been a number of applications using optical tweezers with microfluidics, for example, to sort cells, (MacDonald et al., 2003; Milne et al., 2007; Smith et al., 2007) or to manipulate and measure fluids within microdevices (Leach et al., 2006; Mushfique et al., 2008a,b).

One of the most successful applications of light action in microfluidics has been the "optical stretcher" (Guck et al., 2002, 2001, 2000). An object placed in a dual-beam trap becomes stretched: while the total force acting on the object is zero there is a momentum transfer from the light to the surface as the light passes through the interface acting away from the beam propagation direction. Hence, a *sufficiently soft* object will become stretched in a dual-beam trap. The technique allows one to probe the mechanical properties of the trapped soft particles and, in fact, it has been shown to be a valuable tool for cell mechanics studies, where the deformability of a cell is strictly related to its cytoskeleton. Indeed, optical stretcher has been successfully used as a cell marker for detecting cancerous cells (Guck et al., 2005). Moreover, by integrating the device into a microfluidic system, it has been shown that cells can be analysed in a high throughput manner (Lincoln et al., 2007a,b).

In the field of rheology, optical tweezers have been successfully used with Newtonian fluids for purposes such as determining the fluid viscosity with high accuracy, measuring the hydrodynamic interactions between particles or estimating the wall effect on the Stokes drag coefficient (i.e. Faxén's correction), as reviewed in (Yao et al., 2009a). Conversely, when optical tweezers have been adopted for measuring the viscoelastic properties of complex fluids the results are either limited to the material high frequency response (Atakhorrami et al., 2006; Nijenhuis et al., 2009; Starrs and Bartlett, 2003), discarding the essential information related to long times scales (i.e. low frequency) material behaviour, or supported by low frequency measurements performed by different techniques, such as rotational rheometry (Pesce et al., 2009) or passive video particle tracking microrheology (Nijenhuis et al., 2009), but either without showing a clear overlapping region between the results (Pesce et al., 2009) or even leaving a macroscopic gap of information in the range of frequencies explored (Nijenhuis et al., 2009).

Thus the aim of this book that provides a detailed introduction to a set of microrheology tools developed over the past few years by Tassieri and co-workers (Preece et al., 2011; Tassieri et al., 2010a, 2012; Yao et al., 2009a). Indeed, these methods allow the evaluation of the fluids' linear viscoelastic properties over the *widest* range of experimentally accessible frequencies, without the adoption of *preconceived* models nor the *idealisation* of real measurements.

Part II

Optical Tweezers

Chapter 2

The Nature of Light

R. Mike L. Evans
Department of Applied Mathematics, University of Leeds, Leeds LS2 9JT, UK
r.m.l.evans@leeds.ac.uk

2.1 A Condensed History of Optics

What is light? The question has vexed some of the greatest minds since the dawn of humanity. For most of that history and prehistory, light seemed intangible; a topic of metaphysical debate.

It was only very recently, in the history of academia, that the physical nature of light began to be uncovered. Thanks to the discoveries of Thomas Young, James Clerk Maxwell, and the many physicists and mathematicians on whose intellectual foundations they built, we now know it to be a very tangible phenomenon. In fact, understanding and control of the physical principles that govern it have, in the last few decades, developed to the point that light can be used as a tool with which to exert forces on solid and liquid matter.

While the true nature of light remained a mystery, some of the geometrical principles governing rays of light were discovered by the ancient Greeks. The treatise "Catoptrics" (meaning the study of mirrors), attributed to Euclid, dates to 280 BCE and states the

Microrheology with Optical Tweezers: Principles and Applications
Edited by Manlio Tassieri

ISBN 978-981-4669-18-4 (Hardcover), 978-981-4669-19-1 (eBook)
www.panstanford.com

principle that light travels along straight lines through a single homogeneous medium, such as air. Although this principle, known as the law of rectilinear propagation, appears to be self-apparent, we now know that it holds only in the limit of very large optical apparatus, for which the wave nature of light is not apparent.

Later work by the wonderfully named Hero of Alexandria (first century BCE), also in a book entitled "Catoptrics", included an extraordinarily sophisticated and elegant principle governing the geometry of reflections. Hero discovered the principle that a reflected light ray adopts the shortest possible path from its source, via the mirror, to its destination. This principle of least distance seems at first to be nothing more than a curiosity or a coincidence, since the light ray would require some knowledge of its future destination and of the available choice of paths, in order to select the shortest. Nevertheless, we now understand it to be a profound result of wave physics, due to interference between nearby paths. Moreover, it is the first recorded example of a variational principle, a powerful tool of modern theoretical physics, both in thermodynamics and quantum mechanics.

The principle of least distance was subsequently generalised by the French mathematical genius Pierre de Fermat. In 1657, Fermat showed that the geometry of both reflection and refraction could be derived from a principle of least *time*, if one assumes that light travels more slowly through dense media than through air. This intellectual leap, which we now know to be correct, is quite extraordinary considering that light was not even shown to have a finite speed until the mid-1670s when its speed was first measured.

Prior to that measurement, light was widely believed to travel instantaneously from source to observer. The topic was vigorously debated, with Newton, among others, advocating a finite velocity, but believing it to be *higher* in dense media. For many centuries even, the established belief was in a *negative* velocity. The ancients accepted the assertion of Empedocles (490–430 BCE), that a fire, lit by the goddess Aphrodite, cast rays of light *outwards* from the human eye, to touch the objects gazed upon. Indeed, barring some esoteric wave phenomena, the geometry of ray optics is unaffected by reversing the direction of propagation, so that the Catoptrics of Hero remain valid despite being derived under the Empedoclean assumption.

The correct direction of propagation was first hypothesised by the medieval Arab polymath Alhazen (c. 965–1040 AD), who demonstrated that images could be formed by a camera obscura (a dark room with a hole in one wall), without the intervention of the eye. He made significant discoveries on refraction, lenses, and the formation of images. His understanding of refraction, in his "Kitab al-Manazir" (Book of Optics) was not surpassed until after the renaissance, when geometrical optics received the empirical and mathematical attention of many great thinkers, including Galileo Galilei (1564–1642), Johannes Kepler (1571–1630), Willebrord Snellius (1580–1626), René Descartes (1596–1650) and Carl Friedrich Gauss (1777–1855).

Throughout the long history of optical enquiry, the above scholars uncovered the geometrical principles governing light rays. In other words, they discovered *where* light is, but, with the exception of Alhazen's debunking of Empedocles, they were no closer to discovering *what* light is.

Some clues were found during the seventeenth century. Between 1672 and 1677 the Danish astronomer Ole Rømer made observations which finally proved that light travels with a finite speed. He made a series of accurately timed measurements of the eclipses of Jupiter's moon Io. As Galileo had previously pointed out, the eclipses of Io can be used as a reliable clock because, orbiting Jupiter at a constant speed, it passes through the planet's shadow at regular intervals like the ticking of a clock. Rømer (and also Cassini) noticed a small variation in the rate of Io's "ticking", that related to the relative positions of Earth and Jupiter in their solar orbits. The eclipses of Io appeared to be running behind schedule when Jupiter was far from the Earth, with the lag-time directly proportional to its distance. The discrepancy, he deduced (Fig. 2.1), was due to the finite time taken by light to travel from Io to Earth.

The other significant discovery at around that time was made by Isaac Newton who performed a series of experiments regarding the nature of light (in between his other studies on every subject imaginable, from mechanics and gravitation to alchemy). In 1666, Newton shone a beam of white light through a triangular prism of glass. He famously found that the glass spread the white light into a spectrum of rainbow colours, because each colour was

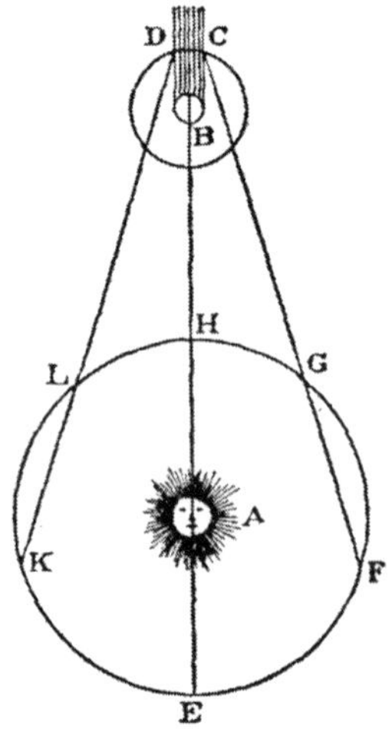

Figure 2.1 Ole Rømer's original drawing (Rømer, 1676), showing the relative positions of Jupiter and Earth and the path travelled by light from one to the other.

refracted through a slightly different angle; a phenomenon known as dispersion.

Newton extended the experiment, redirecting the colourful spreading beam of light so that it once more converged into a narrow spot where the differently coloured light all overlapped. The recombined light was white. So Newton concluded that white light, far from the established perception of being the purest of substances, was a compound thing, composed of elemental coloured components.

Although he had revealed important evidence of light's properties, Newton went on to formulate a theory of the nature of light that we now know to be wrong. Unfortunately, his "corpuscle theory" was highly influential. He supposed that light was composed of coloured particles, called corpuscles (from the Latin for "small body"), that travel at high speed in straight lines. Reflection was then explained by elastic collisions; the corpuscles bounce off shiny surfaces.

Since the development of quantum mechanics in the early twentieth century, a common misconception exists. It seems tempting to conflate the idea of Newton's corpuscles with the modern concept of photons. As we shall discuss in Section 2.6, although photons, the quantum-mechanical "building blocks" of light, are referred to as "particles", the word has a technical meaning within the terminology of quantum mechanics, which is very different from the everyday classical concept of a particle as understood by Newton. The apparent similarity between corpuscles and photons is only superficial, and Newton's corpuscle theory, dominant for over a century, was eventually unequivocally disproved by the pivotal experimental observations of Thomas Young.

The English physicists and linguist Thomas Young (1773–1829) was not the first to suggest that light is a wave phenomenon, but he was the first to demonstrate it rigorously and experimentally. The idea had been advocated, during Newton's lifetime, by the Dutch physicist Christiaan Huygens (1629–1695). His theory described wavefronts of light as geometrical surfaces from which secondary "wavelets" radiate, and determine the position of the next wavefront. It was a geometrical construction (shown in Fig. 2.2) that ensured the direction of propagation was always perpendicular to the wavefronts, and resembled our modern understanding of wave propagation and interference.

Unfortunately, without the modern mathematical tools of partial differential equations, Huygens was unable to formulate a rigorous

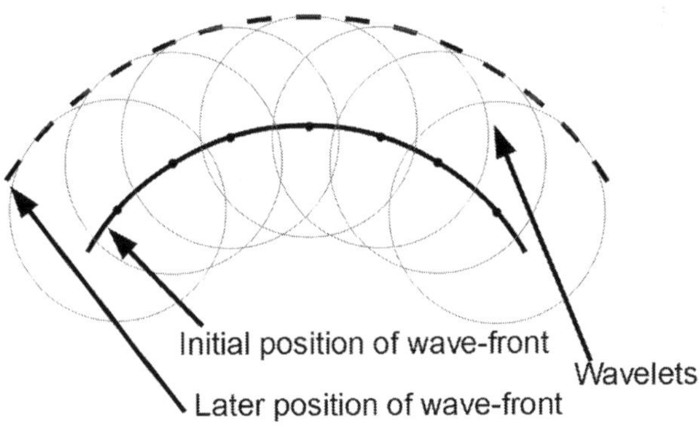

Figure 2.2 Illustration of Huygens's construction, in which each point on a wavefront acts as a source of "wavelets" that combine to produce a new wavefront at a later time. The construction resembles some features of our modern understanding of wave motion by constructive interference of propagators.

description of the wave motion, so Newton was easily able to point out inconsistencies. In particular, in addition to the correct direction of propagation, Huygens's construction also generated spurious backward-propagating waves emanating from every wavefront. Also, it was not understood how to incorporate the recently discovered phenomenon of birefringence into Huygens's theory, nor how waves could travel through a vacuum. So Newton's incorrect corpuscle theory gained dominance until the publication of Young's experimental findings in 1804.

2.2 Wave Physics

The modern age of wave optics began with Young's demonstration of water waves on the surface of a ripple tank. One crucial aspect of wave physics, that differs from the behaviour of particles, is *interference*. When waves from two sources meet, they overlap (superpose) and can either interfere (combine) constructively, creating a wave of larger amplitude than either of the two waves alone, or they can interfere destructively, cancelling out the motion. The type of interference that results depends on the precise timing, or "phases" of the two waves.

In Young's shallow tank of water, he made circular ripples spread out from a pair of dippers, side by side, that rapidly dipped up and down in synchrony. At some particular locations on the water's surface, waves from the two sources arrive *in phase*, with the wave crests coinciding, producing a double-height crest, rapidly followed by two coinciding troughs, producing a trough of double depth. Hence, at these locations of constructive interference, the height of the water surface oscillates with a large amplitude. At other locations, a crest from once source of ripples always coincides with a trough from the other, and vice versa, so that the water remains still.

At any given location, the phase relationship between the two waves depends only on the difference between the two distances to the sources (the dippers). Young's original sketch (Fig. 2.3), where he has shaded the wave troughs at one instant, shows the result very clearly. The locations of large amplitude (constructive interference) lie along lines, on which the *path difference* to the two sources (A

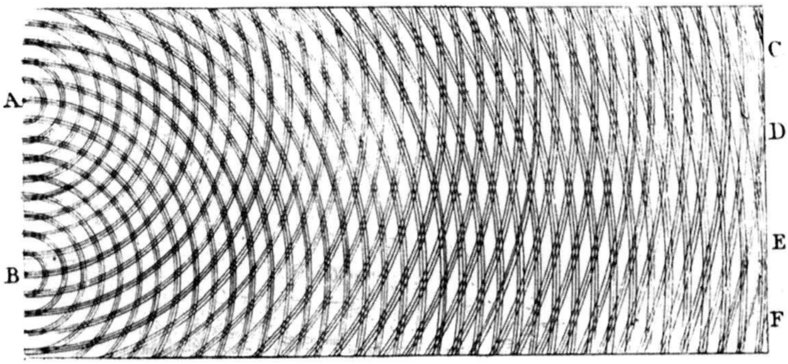

Figure 2.3 Thomas Young's original sketch of his interference experiment in a ripple tank, presented to the Royal Society in 1803. The locations of the wave troughs (low points on the water surface) at one instant are each shaded with three lines. The circular ripples emanate from two oscillated dippers, located at A and B.

and B) remains constant. So the waves' energy is beamed along these directions, separated by lines of still water (marked C to F in Fig. 2.3) where the interference is destructive. Before turning his attention to light, Young demonstrated this previously unknown property of waves.

The other property of waves that Young demonstrated was "diffraction". Diffraction is the spreading out of waves when they pass though an aperture (a hole in a barrier). Considering water waves again, let us imagine still water in a harbour, surrounded by a harbour wall with a narrow opening to the sea. A ship passes by, out at sea, creating a series of straight wavefronts that travel towards the harbour. The waves travel through the harbour mouth and strike the part of the quayside directly opposite the harbour mouth. An angler, sitting on this section of the quayside (at A in Fig. 2.4) and dangling a float into the water, will notice the float bobbing up and down, thus detecting the wave.

Now consider a second angler, stationed further around the quayside, at position B in Fig. 2.4. Before the ship passed, the second angler's fishing float was initially perfectly still, on the surface of the undisturbed water. Will this second angler, fishing in the more sheltered water, also detect waves from the passing of the ship? The answer is yes. Although there is no straight line of sight from

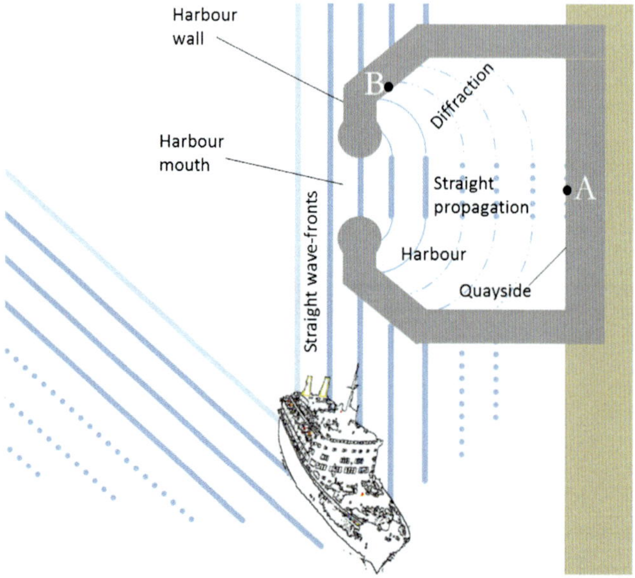

Figure 2.4 Straight wavefronts, from a passing ship, are diffracted at the harbour mouth (an aperture) so that an angler at *B* detects the waves, as well as the angler in the straigh-though direction at *A*.

the source of the waves, though the harbour opening, to point B, common experience tells us that the still water at the edge of the harbour cannot remain perfectly undisturbed while a bow-wave disturbs the water at the nearby harbour-mouth. A disturbance at some point in a pool of still water always spreads out to disturb the whole surface. We can think of the oscillating body of water at the harbour mouth (due to the inbound wave) as playing the role of one of the dippers in Young's ripple tank.

Not only surface water waves, but waves of all kinds share this behaviour of spreading out on passing through an aperture (diffracting). For example, common experience demonstrates that sound waves (pressure waves in the air) also diffract. This allows you to hear people talking in the next room, due to the diffraction of sound waves passing through the open doorway. So your ear is disturbed even when it has no straight "line of sight" from the speaker's mouth.

Diffraction is a property of waves, clearly distinct from the behaviour of particles, which travel in straight lines through openings (respecting Newton's first law of motion). One can imagine particles such as arrows, being fired at a castle by enemy archers. The castle's occupants are liable to be hit if they stand at a window, where they have a direct line of sight to the archers. But they can safely shelter to the side of the window because the arrows will not travel around corners.

Superficially, light appears to travel ballistically, like arrows, because we notice sunlight travelling straight through windows, illuminating a well defined, window-shaped area, and leaving sharp-edged shadows beyond the line of sight. But Young realised that its rectilinear propagation is only approximate, and that the approximation holds only for wide apertures.

As we shall see in the next chapter, plane waves passing through an aperture that is large compared with the wavelength diffract very little, so that the majority of the wave energy continues to propagate in the straight-though direction, or directions that deviate from it by only a small angle. Thus, short-wavelength ripples pass almost unperturbed through a wide harbour mouth, proceeding only straight ahead towards the quayside, whereas a long-wavelength ocean swell affects the whole harbour.

Visible light has a wavelength of order half a micron ($0.5\ \mu\text{m} = 5 \times 10^{-7}$ m), so a window, of order 1 m wide, diffracts the light through an immeasurably small angle. Light travels through such a large aperture with approximately no deviation from the straight-through direction. This limit of rectilinear propagation through apertures that are much larger than the wavelength is known as the ray approximation. On the scale of everyday objects, the ray approximation applies to visible light but not to audible sound, which has wavelengths in the range 17 mm to 17 m.

Thomas Young was the first experimenter to exceed the ray approximation by creating apertures not much wider than the wavelength of visible light. Light, shone through such an aperture, or slit, spreads out, as if the slit itself were a source of light. Two such slits, very closely spaced, behave as two sources of light waves. They play the same role as the dippers (two sources of water waves) in Young's ripple tank. Thus, light waves from these slits interfere, so

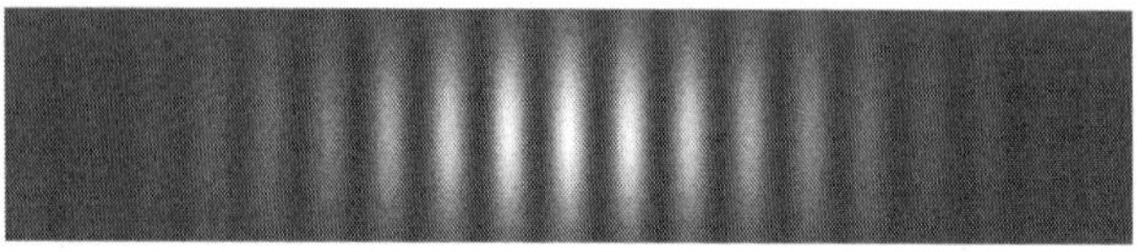

Figure 2.5 The pattern of bright and dark fringes produced by Young's double-slit experiment.

that their power is beamed in a number of distinct directions, exactly like the interference of water waves sketched by Young (Fig. 2.3). When light from the slits illuminated a screen, Young observed a series of light bands separated by dark fringes (Fig. 2.5), thus elegantly demonstrating that Isaac Newton, and almost everyone else since the dawn of time, had misunderstood the nature of light; it is a wave.

Interference experiments, either for light or for water or any other waves, require the sources to remain well synchronised. The ripple-tank dippers, for instance, are rigidly fixed together so that the phase relationship between their waves remains perfectly constant. If the dippers were driven by two imperfectly synchronised motors, the locations of constructive and destructive interference would wander as one motor ran ahead or behind the other by a fraction of a revolution. It would then be difficult to distinguish the shifting interference pattern from a structureless morass of ripples. In the case of light, with a wave frequency of order 10^{15} Hz, the phases of the sources must remain well locked for a vast number of cycles to avoid the interference pattern flickering too quickly to observe. Unsynchronized sources, known as "incoherent" sources, thus produce no observable interference.

So Young's double slit must be illuminated by coherent light; that is, light with orderly waves, so that each wavefront that arrives at one slit is identifiably the same wavefront that arrives at the other slit. The distance over which a wavefront unambiguously retains its identity is called the "coherence length" of the light. Thus Young's slits must be illuminated by light with a coherence length that exceeds the separation of the slits. Figure 2.6 shows examples of coherent and incoherent waves; i.e. waves with long or short coherence lengths.

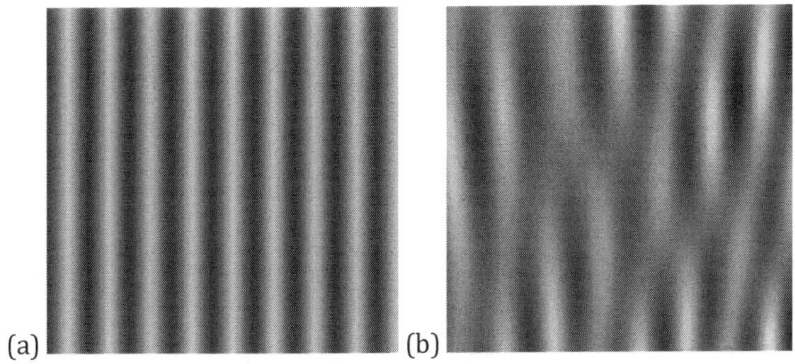

Figure 2.6 Waves indicated by shades of grey, corresponding to the height of the surface in the case of water waves. (a) Coherent waves, with a well defined phase relationship between distant points. (b) Waves with a short coherence length. The phase relationship at different locations is predictable only within a small proximity.

Ordinary light sources tend to be coherent only across distances of a few wavelengths. Until fairly recently, it was a non-trivial task to set up a sufficiently coherent light source to demonstrate interference. A narrow slit illuminated by mercury light was often used. Modern experiments are easily performed using laser light, which typically has a coherence length of order millimetres or more.

Young's revolutionary experiments, at the end of the eighteenth century, went a long way toward answering the age-old question: what is light? He had shown unequivocally that it is a wave phenomenon. But a wave of what? While water waves transport energy by perturbing the height of the water's surface, and sound transports energy by varying the pressure of a gas, the substance of light remained unknown for a further sixty years.

2.3 Electromagnetism

The great Scottish mathematical physicist James Clerk Maxwell (1831–1879) finally had the crucial insights into the electromagnetic nature of light waves. He developed a unified theory of electricity and magnetism out of the collected observations of a number of scientists. Many textbooks are now devoted to deriving and explaining the principles of electromagnetism in full detail. A brief overview of those principles is given below, because a knowledge of them is necessary for our two purposes of understanding the true nature of light, and using it to control matter in optical tweezers.

2.3.1 *Fields*

Maxwell showed how a disparate set of experimental results, involving copper wire, magnets and voltaic cells (batteries), revealed the laws of nature governing the interactions of fundamental fields, and could be expressed in four beautifully simple equations.

The fundamental subjects of his theory are the electric and magnetic fields. The term "field" means a function of position, so these are local quantities, whose strength and direction vary from place to place. They are both vector quantities, meaning that they have both a magnitude (or strength) and a direction. In SI units, the magnetic field, denoted $\mathbf{B}$, is measured in tesla (T) and the electric field $\mathbf{E}$, in volts per metre, V m^{-1}. The integral of electric field along the length of some path (e.g. along a wire) gives the voltage (properly called the electric potential difference) measured across its ends.

Both phenomena can be felt as a real visceral presence. The magnetic field can be felt by holding a permanent magnet in each hand and trying to push like-poles together, or moving them past each other. The electric field is felt when an eletrostatically charged balloon (charged up by rubbing against clothing) is held close to ones arm, because it exerts a force on the charges within all materials; in this case in the small hairs on the arm. In fact the strength of the electric field $\mathbf{E}$ is *defined* as the electrostatic force (measured in Newtons, N) on a test charge introduced into that field, per unit of its charge (measured in Coulombs C).

2.3.2 *Coulomb's Law and Gauss's Law*

As well as applying a force to a charge, the electric field is produced by charges. As the French physicist Coulomb (1736–1806) discovered (Coulomb, 1785), each electrostatic charge (such as the electrostatic charge on a rubbed balloon) is surrounded by an electric field, oriented around it like the spines around a hedgehog. The field is strongest close to the charge, and decays away as the inverse square of the distance from it. Also, the greater the amount of charge of magnitude Q, the stronger the field. So for the field due to a point charge Q at the origin, at a distance r from it in the direction of the radial unit vector $\hat{\mathbf{r}}$, Coulomb's law can be expressed mathematically as

$$\mathbf{E} = \frac{1}{4\pi\varepsilon_0}\frac{Q}{r^2}\hat{\mathbf{r}}. \tag{2.1}$$

The value ε_0 in the constant of proportionality is known as the *permittivity of free space*. Due to the definition of electric field, the force between two charges at separation r is given by multiplying Eq. 2.1 by the magnitude of the second charge. So,

> Coulomb's law states that the electrostatic force between two point charges is proportional to the magnitude of each charge, inversely proportional to the square of their separation, and directed along the line joining them. The force is attractive for charges of opposite sign and repulsive for charges of like sign.

As the German mathematician Carl Friedrich Gauss (1777–1855) discovered, if we imagine surrounding a charge by a closed surface S, as in Fig. 2.7, then the total amount of electric field crossing that surface (integrated over its area) does not depend on the size or shape of the surface. This is because a surface placed close to the charge must have a small area, but encounters a strong field, whereas, far from the charge, the field is weak, but is integrated over a large area. The area of the surface is proportional to the square of its radius, while the electric field is proportional to the inverse square, leading to a neat cancellation. The surface integral depends only on the amount of charge enclosed, thus

$$\oint\!\!\oint_S \mathbf{E}\cdot d\mathbf{S} = \frac{Q}{\varepsilon_0} \tag{2.2}$$

where $d\mathbf{S}$ is a vector with magnitude of an area element and direction normal to the surface (so that the scalar dot product measures only the component of $\mathbf{E}$ that crosses the surface).

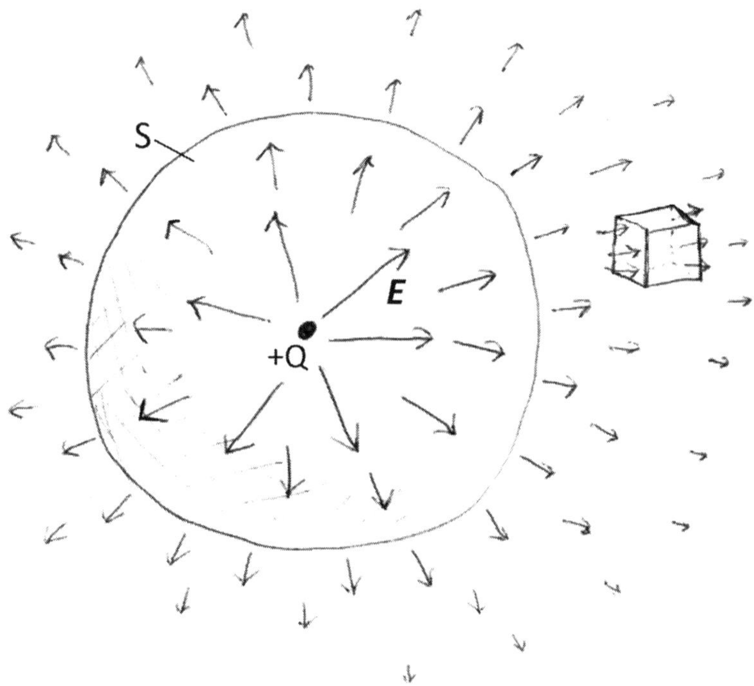

Figure 2.7 A hypothetical surface S surrounding an electrostatic charge Q that is the source of electric field E. A smaller box is also shown, enclosing no charge, so that equal amounts (fluxes) of electric field enter and leave the box.

Although Gauss's law (Eq. 2.2) involves a large-scale quantity, spread over a surface, it can be mathematically transformed into a relationship between local quantities, defined at each point in space. Conceptually, this is done by shrinking the surface down to an infinitesimal size. We shall not reproduce the derivation here, as it can be found in textbooks on electromagnetism, and introduces no new physical observations or assumptions. Purely by mathematical manipulation, one can re-cast Eq. 2.2 as

$$\text{div}\,\mathbf{E} = \frac{\rho}{\varepsilon_0} \tag{2.3}$$

where ρ is the electrostatic charge density (charge per unit volume), defined at each point in space. At most locations in Fig. 2.7, $\rho = 0$, but it is large at the location of the charge, equal to Q

divided by the volume of the charge. Hence, in the limit of a point charge at position vector $\mathbf{r}_0$, the charge density becomes infinite but integrable, $\rho(\mathbf{r}) = Q\,\delta^{(3)}(\mathbf{r} - \mathbf{r}_0)$ where $\delta^{(3)}(\mathbf{r})$ is a three-dimensional Dirac delta function. The *divergence* of the vector field **E**, appearing in Eq. 2.3, can be expressed using a scalar product with the vector differential operator del, ∇, which has Cartesian components $\nabla = (\partial/\partial x, \partial/\partial y, \partial/\partial z)$, thus

$$\mathrm{div}\,\mathbf{E} \equiv \nabla \cdot \mathbf{E} = \frac{\partial E_x}{\partial x} + \frac{\partial E_y}{\partial y} + \frac{\partial E_z}{\partial z} \qquad (2.4)$$

for Cartesian components of **E** given by $\mathbf{E} = (E_x, E_y, E_z)$.

Physically, the quantity div **E** measures how much the vector field spreads out at each point. Equation 2.3 shows that electric field only diverges from electrostatic charges. In a location with no charge density, such as at the small (assumed to be infinitesimal) box in Fig. 2.7, there is no divergence. Although the directions of **E** are splaying outwards, its magnitude is diminishing in the radial direction, so that the integrated amount (flux) of electric field entering the infinitesimal box is equal to the amount leaving.

Equation 2.3 is the result of the combined work of many physicists and mathematicians, including Coulomb, Gauss, Maxwell and Heaviside (1850–1925). But Maxwell realised its significance as one of a handful of equation (now known as Maxwell's Equations, for which Eq. 2.3 is the first) that constitute the laws of nature governing the electric and magnetic fields. He also revealed its relevance to the nature of light.

The second of Maxwell's Equations is almost identical to the first. It expresses the fact that, in empty space, magnetic fields behave in the same way as electric fields, becoming weaker as they spread out. It also embodies the observation that there is no magnetic equivalent of an electrostatic charge. Magnetic north and south poles always come in pairs, and no magnetic monopole has ever been observed. Hence, the magnetic monopole density is zero on the right-hand side of Maxwell's second equation,

$$\mathrm{div}\,\mathbf{B} = 0. \qquad (2.5)$$

2.3.3 *Faraday's Law*

The results of the English physicist Michael Faraday (1791–1867) and French physicist André-Marie Ampère (1775–1836) were central to Maxwell's theory. Faraday had discovered the phenomenon of electromagnetic induction, which is the basis of all present-day power stations. The phenomenon is easily reproduced, in table-top experiments at home or school, by connecting a coil or loop of wire to a light bulb or voltmeter, as in Fig. 2.8. Waving a magnet past the coil generates a voltage (properly termed "electromotive force" or EMF, also called "electric potential" and measured in volts V) across its ends, that can be measured by the voltmeter and used to drive a current through the light bulb.

Faraday found that the voltage generated was proportional to the rate of change of the magnetic flux passing through the loop of

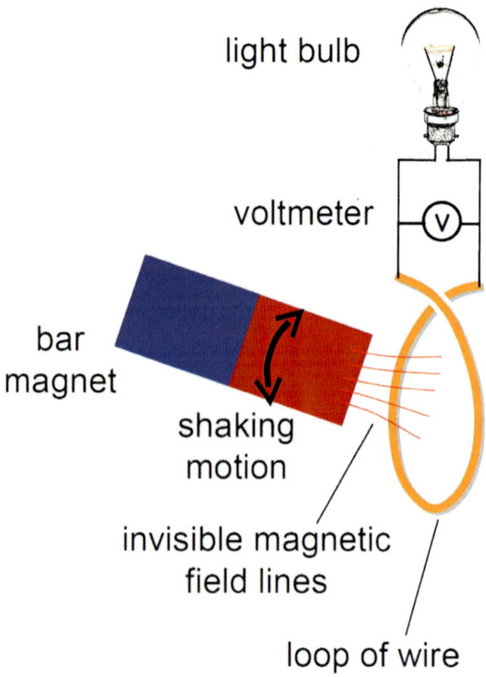

Figure 2.8 A simple circuit to demonstrate electromagnetic induction, as described by Faraday's law.

wire. So electricity can be generated either by bringing one pole of a magnet close to the loop, or by taking it away. This is expressed in Faraday's law.

> Faraday's law states that the induced electromotive force V in any closed circuit is equal to the negative rate of change of the magnetic flux Φ through the circuit.

That is,

$$V = -\frac{\mathrm{d}\Phi}{\mathrm{d}t} \tag{2.6}$$

where t represents time, so that rate of change is expressed as the time derivative $\mathrm{d}/\mathrm{d}t$. The negative sign arises from the standard convention chosen for defining the directions of increasing electric potential and positive magnetic field. Magnetic flux Φ can be loosely understood as the amount of magnetism emanating from the magnet. More precisely, it is the total amount of magnetic field $\mathbf{B}$ integrated over the area of the loop, so flux has units T m^2.

Maxwell realised that Faraday's law, which is a relationship between two gross quantities—the total amount of flux passing through the entire loop, and the overall voltage measured across its ends—could be restated in terms of local quantities, giving insight into the physical process occurring at each point in space. The changing magnetic field (due to the waving magnet) induces an electric field at each point around the loop of wire. The total of all this electric field, integrated around the loop, is the voltage measured across its ends.

Maxwell derived mathematically, from Eq. 2.6, that, at each point in space, the rate of change of magnetic field generates a local non-uniformity of the electric field, making it twist. In modern notation, this is expressed in the neat mathematical relationship

$$\text{curl}\,\mathbf{E} = -\frac{\partial \mathbf{B}}{\partial t} \tag{2.7}$$

where the curl of the electric field is the vector (cross) product of the del operator with the electric field thus,

$$\text{curl}\,\mathbf{E} \equiv \nabla \times \mathbf{E} = \left(\frac{\partial E_z}{\partial y} - \frac{\partial E_y}{\partial z}, \frac{\partial E_x}{\partial z} - \frac{\partial E_z}{\partial x}, \frac{\partial E_y}{\partial x} - \frac{\partial E_x}{\partial y} \right).$$

Maxwell had the great insight that Eq. 2.7 remains true even in the absence of the wire or any of the experimental equipment in Fig. 2.6. It is not just a feature of electrical circuits, but a law of nature governing the fundamental fields **E** and **B** that are properties of space.

2.3.4 *Ampère's Law and Displacement Current*

Maxwell derived another law of nature governing the **E** and **B** fields by considering the results of another experiment on electrical circuits—that of Ampère. Whereas Faraday had found that magnetism can generate electricity, Ampère discovered that electricity generates magnetism. It is another demonstration familiar to modern schoolchildren, that sending an electric current (a movement of electric charge) through a coil of wire creates an electromagnet that can turn a compass needle or pick up paper-clips. Ampère measured the strength of the magnetic force produced by a given current. His measurements revealed a simple relationship between the amount of electric current flowing along a wire and the strength of the magnetic field that it created.

A single wire, carrying a current I creates magnetic field oriented in a direction that circles the wire. The strength of this field respects and elegant principle, for which we must imagine constructing a hypothetical loop around the wire (see Fig. 2.9). Integrating the field around any such closed loop that is threaded by the wire gives a result that is proportional to the current, but does not depend on the length of the loop or how far it is from the wire. This is because a loop drawn very far from the wire must be very long, but the field that is integrated along it is very weak, because the field strength decays with distance from the wire. This principle is expressed in Ampère's law.

> Ampère's law states that the circulation of magnetic field around any closed loop is proportional to the electric current threading that loop.

The constant of proportionality is called the *permeability of free space*, μ_0. Here, the term "circulation" has a specific mathematical

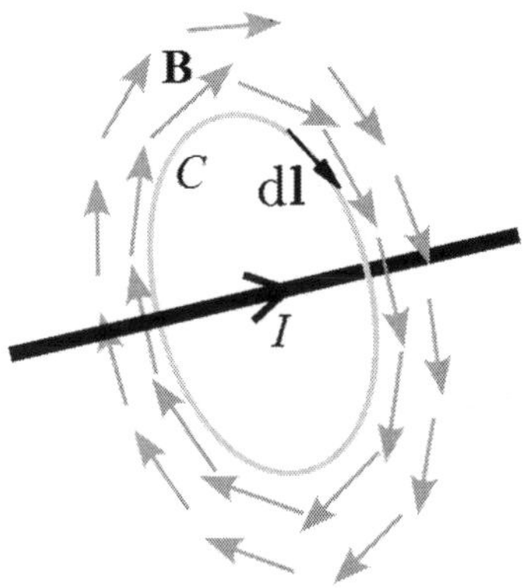

Figure 2.9 The magnetic field **B** around a wire carrying an electric current I. Also shown is the hypothetical construction (used in Ampère's law) of a closed curve C (with vector line element **dl** around the wire.

meaning. It is the integral, around a closed loop, of the component of the vector field **B** oriented along the loop (in the direction of integration). So Ampère's law is expressed mathematically as

$$\oint_C \mathbf{B} \cdot \mathrm{d}\mathbf{l} = \mu_0 I \tag{2.8}$$

where I is the total current threading the closed curve C.

One of Maxwell's greatest strokes of genius was to realise that Ampère's law could not be the whole truth, since it contained and inconsistency, and furthermore, with no new experimental data, to derive a previously unknown law of nature that would correct the inconsistency. He considered what would happen to Ampère's result if there was a break in the wire, across which no current could flow. One might suppose that the current would have to vanish all along the wire, but that is not necessarily that case. It is possible for the current to continue flowing in the wire, leading to a build-up of positive charge on one side of the break and of negative charge on the other. In fact, that is exactly what happens in a capacitor, an electrical circuit component with the capacity to store large amounts of opposite charges, spread across two large plates (typically rolled up into a small space) separated by a narrow gap—a break in the circuit.

Maxwell considered the effect of such a break in the wire, located near the centre of the hypothetical closed curve C of Ampère's law (see Fig. 2.9). The current I is zero in the narrow gap, and finite elsewhere along the wire. The hypothetical curve need not be a perfect circle, or even planar. So, is the curve C threaded by a finite current or not? The value of the current I becomes ambiguous. Ampère's law makes two conflicting predictions for the value of the circulation of magnetic field around the curve, depending on whether one considers that the curve is threaded by the wire or by the gap. In fact, a mathematical identity, beyond the scope of this introduction, exists, proving that it is impossible to distinguish between the two cases.

Reasoning that the magnetic field cannot be ambiguous, as it is a physical object with a measurable value, Maxwell, understood that Ampère's law must be incomplete. He realised that, not only does the current I generate a magnetic circulation; the gap must do so too. The only feature distinguishing this gap from any other empty space (that does not spontaneously generate magnetic circulation) is the fact that it is close to an increasing build-up of charge. So the electric field within the gap is increasing. Maxwell postulated (correctly) that a changing electric field plays the same role as an electric current, in generating magnetic circulation.

A calculation, involving conservation of charge at the wire's ends, shows that the quantity in the gap corresponding to (i.e. equal to) the current in the wire is

$$\int_S \varepsilon_0 \frac{\partial \mathbf{E}}{\partial t} \cdot d\mathbf{S} = I \tag{2.9}$$

where the integral is over a surface S bounded by curve C and the constant of proportionality ε_0, known as the *permittivity of free space*, relates a quantity of electric charge to the strength of electric field around it. The quantity on the left-hand side of Eq. 2.9 is called the *displacement current* and it is added to the right-hand side of Eq. 2.8 to correct Ampère's law.

As well as discovering this new mechanism for generating magnetic fields, Maxwell re-cast the integral equation of Ampère's law into an elegant relationship between locally-defined quantities, as he had for Faraday's law (see above). Let us define a local electric current density $\mathbf{J}$ at each point within the wire, as the current per unit area crossing any surface. So the total current I is given by integrating $\mathbf{J}$ across the cross-sectional area of the wire. Finally, Maxwell's corrected form of Ampère's law leads (after some mathematical manipulations of Eqs. 2.8 and 2.9) to a neat relationship between the fields,

$$\text{curl}\,\mathbf{B} = \mu_0 \left(\mathbf{J} + \varepsilon_0 \frac{\partial \mathbf{E}}{\partial t} \right). \tag{2.10}$$

2.3.5 *Electromagnetism and Light*

The set of four Eqs. 2.3, 2.5, 2.7 and 2.10 are Maxwell's equations, which govern the behaviour of electric and magnetic fields. We see that the laws of nature treat the fields **E** and **B** in almost identical ways. In fact the only difference is the absence of magnetic charges and currents. Maxwell's discovery of this elegant electromagnetic symmetry can reasonably be regarded as the first unification of fundamental forces.

As we have seen, Maxwell's most substantial contribution to the physics of these equations was his discovery of displacement currents that generate magnetism from a changing electric field. As Maxwell realised, the significance of that discovery is immense, as we shall now see.

Consider the behaviour of the electromagnetic fields in an empty volume of space, free from any charges or electric currents, so that ρ and $\mathbf{J}$ both vanish. Now Eq. 2.7 says that electric field variations are created by a changing magnetic field, and Eq. 2.10 says that magnetic field variations are created by a changing electric field. The result is that creating a disturbance in either field (e.g. by waving a magnet) will generate a disturbance in the other field which, in turn, will generate further disturbance in the first field. So the disturbance will continue *ad infinitum*, propagating as a wave of electromagnetic variation.

Furthermore Maxwell was able to calculate the speed of his proposed electromagnetic waves (as we shall do below). Clearly, it must be some function of the only parameters in Maxwell's equations, ε_0 and μ_0, the permittivity and permeability of free space. Measurements of both quantities already existed in the scientific literature, albeit fairly inaccurate ones. Using those values, Maxwell predicted the speed of electromagnetic waves to be approximately equal to the known speed of light. He concluded:

> "Hence the velocity of light deduced from experiment agrees sufficiently well with the value of v deduced from the only set of experiments we as yet possess...
>
> ...The agreement of the results seems to show that light and magnetism are affections of the same substance, and that light is an electromagnetic disturbance propagated through the field according to electromagnetic laws." (Maxwell, 1864)

The nature of light, pondered throughout all of history, had finally been revealed by James Clerk Maxwell. It is a wave in the electromagnetic fields.

Using modern vector calculus, the derivation and solution of the wave equation from Maxwell's equations is quite straightforward. We take the curl of both sides of Eq. 2.7, and re-write the left-hand side using the vector operator identity

$$\nabla \times \nabla \times \mathbf{E} \equiv \nabla(\nabla \cdot \mathbf{E}) - \nabla^2 \mathbf{E},$$

and the right-hand side using commutation of operators

$$\nabla \times \frac{\partial \mathbf{B}}{\partial t} \equiv \frac{\partial}{\partial t} \nabla \times \mathbf{B}$$

then substitute for $\nabla \times \mathbf{B}$ from Eq. 2.10 with $\mathbf{J} = \mathbf{0}$ and substitute for $\nabla \cdot \mathbf{E}$ from Eq. 2.3 with $\rho = 0$. The resulting equation of motion for the electric field is a partial differential equation of a standard form, known as the wave equation,

$$\frac{\partial^2 \mathbf{E}}{\partial t^2} = c^2 \nabla^2 \mathbf{E} \tag{2.11}$$

with a coefficient $c^2 = 1/(\varepsilon_0 \mu_0)$. The wave equation has a travelling-wave solution, for waves travelling at constant velocity c in any direction. For example, if the field varies only along the direction of the x-axis, then the wave equation simplifies to

$$\frac{\partial^2 \mathbf{E}}{\partial t^2} = c^2 \frac{\partial^2 \mathbf{E}}{\partial x^2}$$

and is solved, for instance, by $\mathbf{E} = \mathbf{E}_0 \sin(k\,[x - ct])$ for constant $\mathbf{E}_0$ and k, as can easily be checked by substitution.

So we have derived the result that waves in the electric field travel through a vacuum at a speed $c = 1/\sqrt{\varepsilon_0\mu_0}$, which is the speed of light. An identical result for the magnetic field can be found by taking the curl of Eq. 2.10. The wave in the electric and magnetic field are not independent, but travel together, with the fields oriented at right angles to each other and to the direction of propagation.

Like any other wave, this electromagnetic wave carries with it both energy and momentum. Furthermore, because it generates electric fields, it can even be used to exert force on the charges within matter, as shall be shown in the remaining chapters of this book.

By 1887, Maxwell's theory was vindicated, when German physicist Heinrich Hertz (1857–1894) experimentally confirmed his prediction, that electromagnetic waves, of much lower frequency than visible light, can be created by electrical circuits. Indeed, the electromagnetic fields can be made to oscillate at any frequency. The resulting waves at the lowest frequencies, and up to around 10^{10} Hz, are called radio waves, such as Hertz produced and detected. At higher frequencies, the waves have shorter wavelength (since wavelength $\lambda = c/\nu$ for frequency ν) and, between around 10^{10} Hz and 10^{12} Hz, the waves, of order millimetres in length, are termed microwaves, although they are no different in nature from radio waves, or any other band of the electromagnetic spectrum. Further increasing the frequency, the waves are named, in turn, infra-red, visible light, ultra-violet, x-rays, and finally, for all higher frequencies, gamma rays. Since Hertz's pioneering experiments, all of these bands of the electromagnetic spectrum have subsequently been detected, each interacting most readily with different aspects of matter. The full spectrum is summarized in Fig. 2.10.

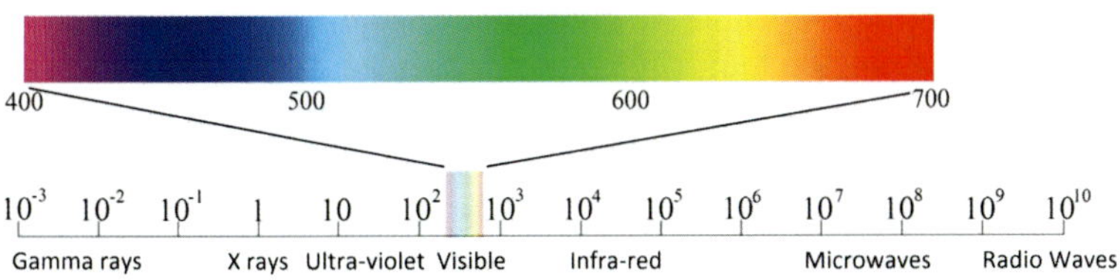

Figure 2.10 The electromagnetic spectrum, with wavelengths in nanometres.

2.4 Interaction of Light with Matter

The wave equation (Eq. 2.11) is derived above for electromagnetic fields in a vacuum. Also, the form of Maxwell's equations, as quoted above (Eqs. 2.3, 2.5, 2.7 and 2.10), assumes that, in between any charges that are present, there is no other matter; only empty space. This is an accurate assumption, given that all matter (at least, the non-exotic forms of matter capable of interacting with light), is composed of charged particles: positively charged protons in the atomic nuclei, surrounded by negatively charged electrons (each with a charge equal and opposite to that of a proton).

Using Maxwell's equations, in this form, to describe the interactions between light and matter, requires us to give a microscopically detailed description of the charge density field ρ (and its current **J**), including the subatomic densities of positive and negative charge. Given that information, Maxwell's equations yield the resulting **E** and **B** fields. In turn, those fields apply forces to the charged particles, so we must also somehow model the ways in which the charge-density and current evolve, in response to the forces imposed by the **E** and **B** fields.

Let us first consider the propagation of light through a transparent medium such as glass or water. These materials are composed of electrically neutral molecules, in each of which the numbers of positive and negative charges balance. Within each molecule, the low-mass electrons form a diffuse negatively charged cloud, anchored by electrostatic attraction (respecting Coulomb's law—see Section 2.3.2) to the much more massive, positively charged nuclei. The precise shape of the electron cloud is governed by quantum-mechanical laws of nature that are beyond the scope of this book. But their response to a perturbing force is easy to understand.

For example, consider a molecule, initially in its equilibrium state, momentarily subjected to some external force, such as an electric field, which displaces the electron cloud a little to the left. After the application of this force, the molecule has a local charge imbalance, with its left-hand side more negatively charged, and its right-hand side therefore more positively charged (less negative) than in the equilibrium state. So, compared with the equilibrium state, the negatively charged electrons within this perturbed molecule experience an extra repulsion from the enhanced electron density on the left, and an extra attraction from the more positive region on the right. The result, then, is a restoring force, tending to accelerate the electron cloud back towards its equilibrium position. For small perturbations, the magnitude of the net restoring force is proportional to the average displacement of the cloud. Hence, the centre of mass of the electron cloud behaves exactly like a mass on a spring tethered to the nuclei, where the spring obeys Hooke's law (that the force in the spring is proportional to its extension, and opposes the extension).

So we see that, while the equilibrium electronic structure of a molecule is difficult to calculate, requiring advanced quantum mechanics, its net response to a small perturbation is just that of a mass on a spring: a simple harmonic oscillator. When the molecule experiences the oscillating electromagnetic field of a light wave, its electron cloud behaves as a simple harmonic oscillator driven by an oscillating external force.

The *classical* driven oscillator is a well studied topic of elementary physics. It responds to the driving force by oscillating with a large amplitude if the driving oscillation matches the oscillator's natural frequency—the phenomenon of *resonance*—and small amplitude for frequencies far from resonance. In the quantum mechanical (*non-classical*) case that applies to very small (e.g. electronic) masses, the response of a simple harmonic oscillator can be more complex close to resonance. But, at much lower frequencies, its response is similar to the classical case. Slowly forcing a small mass on a spring causes it to move at the same (forcing) frequency. Any finite mass will not quite "keep up with" the force; i.e. its maximum displacement will occur a little after the maximum force was applied. The larger the mass, the more the phase of its displacement lags behind the force.

The oscillating electron cloud within a molecule, in turn, has an effect on the electromagnetic fields. By Coulomb's law, each electron is surrounded by a spherically symmetric electric field, centred on the charge. Shaking the charge disturbs that field. As we saw in Section 2.3.5, a disturbance in either the electric or magnetic field propagates outwards as an electromagnetic wave. So the oscillating charge distribution in the molecule re-radiates light, thus giving up the kinetic and potential energy given to it by the incoming light. But the phase of the new light lags a little behind the incoming wave because of the oscillating electron's finite mass.

The result of combining the incident and re-emitted waves is that progress of light through the transparent medium proceeds without any net loss of energy, but it is *delayed* compared with its progress through a vacuum. Thus, light travels more slowly through a dense medium. As discussed in Chapter 2, this fact allows us to design optical glassware to control light's propagation.

A further consequence of the electrons' behaviour as simple harmonic oscillators is to delay high-frequency waves more than low-frequency waves. So a mixture of light waves with a variety frequencies becomes dispersed, with each frequency (colour) of light propagating slightly differently though the refractive medium, as discovered by Newton.

It would be more convenient to be able to calculate the behaviour of electromagnetic fields, passing through materials, without having to model the detailed charge distribution within every constituent atom. As we have seen, the net result of making small electrical perturbations to the molecules in ordinary materials is only to modify the electromagnetic field by amounts proportional to their amplitudes. This linear (i.e. proportional) response of matter that is overall electrically neutral (containing equal amounts of positive and negative charge bound together) can be summarized by a simple modification of Maxwell's equations. On length scales that are large compared with the constituent molecules, each material can be treated as an electrically neutral continuum characterised by some modified (and frequency-dependent) effective values of the parameters ε_0 and μ_0 governing the propagation of electric and magnetic fields. They are scaled by material-dependent parameters known, respectively, as the *relative permittivity* (or *dielectric constant*) ε_r and *relative permeability* μ_r of the material. In the resulting coarse-grained version of Maxwell's equations, only *free* charges and currents (as opposed to those within the structure of the neutral material) need to be included. Thus, in terms of the free charge density ρ_f and free current density $\mathbf{J}_f$, the resulting coarse-grained, or macroscopic versions of Maxwell's Equations are

$$\text{div}\,\mathbf{D} = \rho_f \tag{2.12}$$

$$\text{div}\,\mathbf{B} = 0 \tag{2.13}$$

$$\text{curl}\,\mathbf{E} = -\frac{\partial \mathbf{B}}{\partial t} \tag{2.14}$$

$$\text{curl}\,\mathbf{H} = \left(\mathbf{J}_f + \frac{\partial \mathbf{D}}{\partial t}\right). \tag{2.15}$$

where the re-scaled fields are known as the displacement field $\mathbf{D} = \varepsilon_r \varepsilon_0 \mathbf{E}$ and the magnetizing field $\mathbf{H} = \mu_r \mu_0 \mathbf{B}$.

2.5 Interaction of Light with Metals

Having discussed the interaction of light with materials (known as *dielectrics*) that are transparent to electromagnetic fields, at least for some range of frequencies, let us move on, to consider the interaction of light with *metals*. Metals are electrically conductive because, as well as containing some electrons that are strongly anchored to their atomic nuclei, a significant portion of a metal's population of electrons is highly mobile, able to flow almost unimpeded across the entire length of the sample, in response to applied electromagnetic fields.

Without going into mathematical detail, it is possible to appreciate the physics of light's interaction with a metallic surface by considering a simple experiment. It is an informative practical demonstration that is very straightforward to reproduce, and we encourage the reader to do so. It simply involves dropping a strong permanent magnet though a straight section of copper pipe, held vertically. Ideally, the magnet will not touch the sides as is falls. Since copper is not ferromagnetic, the magnet is not attracted to it, and yet the magnet is observed to fall very slowly though the pipe, taking much longer than it would to cover the same distance in free-fall.

The explanation for this phenomenon is found by considering the cylindrical pipe to be a horizontal loop of wire, elongated in the vertical direction. By Faraday's law (see Section 2.3.3), the movement of the magnet generates an electric current, circulating around the loop, as shown in Fig. 2.11. That electric current, in

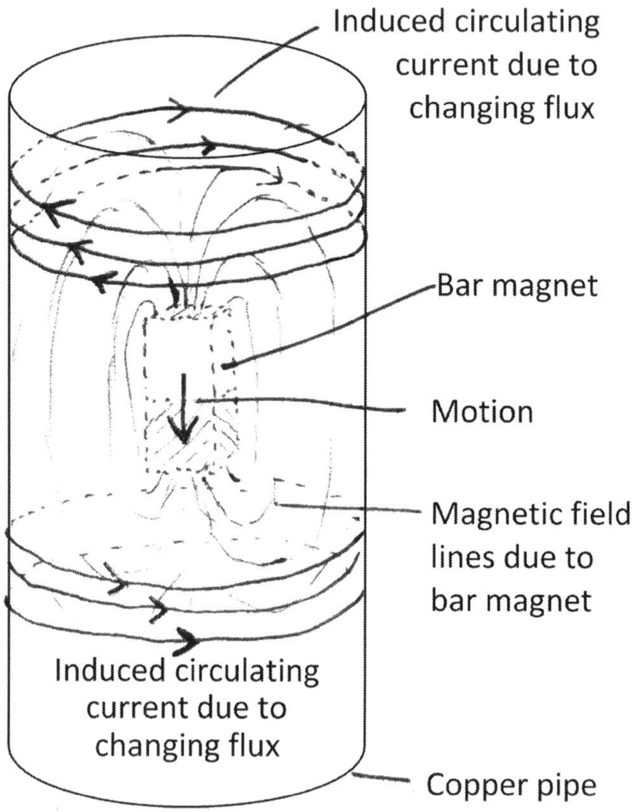

Figure 2.11 A permanent magnet, falling through a vertical section of copper pipe, generated a circulating electric current, by Faraday's law.

turn, generates a magnetic field, oriented vertically, according to Ampère's law (see Section 2.3.4).

The direction (up or down) of that resulting field is given by a neat argument due to the German-Russian physicist Heinrich Lenz (1804–1865). He realised that the induced magnetic field *cannot* be oriented so as to attract the permanent magnet downwards, as that would further accelerate it (in addition to gravitational acceleration), causing it to exit the pipe with more kinetic energy than it would have gained in free-fall. That enhancement to the kinetic energy, with no power source, would violate the principle of conservation of energy. Hence, the resulting magnetic field has the opposite orientation, as stated by Lenz's law.

> Lenz's law states that an induced electromotive force gives rise to a current whose magnetic field opposes the change in magnetic flux that generated it.

Although both electric and magnetic fields are present in a light wave, Lenz's law provides an intuitive understanding of the response of a metallic surface to an incident electromagnetic wave. The electromagnetic oscillations of the wave generate oscillating electric currents in the conductive surface, with a phase and orientation opposing the incident field-components parallel to the surface. Thus, the induced oscillating currents generate an outgoing (reflected) wave. Hence, metals are reflective precisely because they conduct electricity.

2.6 Photons and Lasers

We have seen that the electromagnetic wave-physics, discovered by Young and Maxwell, elegantly explains the propagation of light, and its interference, diffraction, refraction, dispersion and reflection. It is a very accurate theory, able to describe every aspect of the behaviour that had been discovered up to the early twentieth century. However, as explained respectively by the German physicists Albert Einstein (1879–1955) and Max Planck (1858–1947), the quantum-mechanical behaviour of light in the photoelectric effect and in black-body radiation require Maxwell's theory to be modified.

Einstein and Planck demonstrated that electromagnetic waves cannot be excited with continuously variable amounts of energy. Instead, energy is exchanged with the electromagnetic fields in minute discrete quanta, called *photons*. It is important to understand that photons do not correspond to the discrete ballistic corpuscles hypothesized by Newton, and that the existence of photons does *not* overturn Maxwell's theory, but only modifies it, in a way that is perceptible only at the smallest energy-scales.

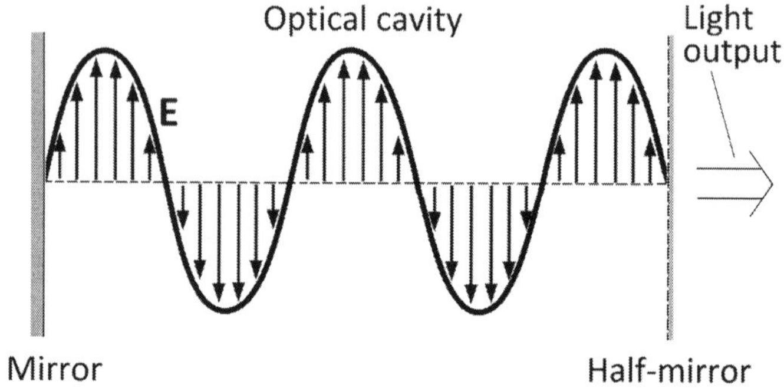

Figure 2.12 An electromagnetic standing wave in the optical cavity of a laser, between two parallel mirrors.

Quantum electrodynamics (QED), the quantum-mechanical version of Maxwell's theory, is beyond the scope of this book, but it is worth pursuing the relationship between photons and electromagnetic waves. The relationship is most clearly demonstrated in the workings of a laser, the most important component of optical tweezers. At the heart of a laser are two parallel mirrors, facing one another, as in Fig. 2.12, and light waves are trapped between them. One of the mirrors is partially transparent, so that some light can be emitted by the laser.

Boundary conditions, imposed (by Lenz's law—see Section 2.5) on the fields at the mirrored surfaces compel the intervening cavity, of length L to contain an integer number of half-wavelengths of trapped light (see Fig. 2.12). So the laser can support only a discrete set of possible wavelengths (and therefore frequencies) of light. In other words, the optical cavity contains electromagnetic standing waves, equivalent to the acoustic standing waves of air pressure confined to an organ pipe, or the harmonics of a guitar string.

Typically a laser is designed to use a single *mode* of the cavity, i.e. a particular standing wave with a specific number of half-wavelengths. Like the displacements of a guitar string, the values of the $\mathbf{E}$ and $\mathbf{B}$ fields at each location in the cavity oscillate sinusoidally with time, because this electromagnetic mode of the cavity behaves as one simple harmonic oscillator. Note that, far from being overturned by quantum mechanics, Maxwell's equations are used thus far to describe the fields in the optical cavity of the laser.

The quantum mechanical laws of nature impose a constraint on the energy (and therefore the amplitude) of excitations of any simple harmonic oscillator. If the oscillator has a natural frequency ν (in this case, set by the length of the cavity L), then the oscillator can only possess certain discrete values of energy, specifically any positive integer multiple of $h\nu$, where the universal constant h is Planck's constant. Hence, in the case of the laser, where the simple harmonic oscillator is composed of electromagnetic fields, the stored light energy can only be increased (by a power supply) or decreased (by light leaving through the semi-reflective mirror) in discrete amounts, known as photons. Note then that, although photons are "particles" in the terminology of quantum mechanics, a quantum mechanical particle is a quantized excitation of a wave.

Quantum electrodynamics is able to describe the quantum-mechanical behaviour of all electromagnetic fields, not only those trapped in a laser. Conceptually, a laser can be turned into an infinite volume of free space, simply by moving its mirrors apart, making it infinitely long. The argument, given above, for the discretization of electromagnetic wave energy into quanta of size $h\nu$, did not rely on any particular value of L.

The thermodynamics involved in injecting energy into the particular simple harmonic mode of the laser's optical cavity (known as "pumping") is quite complex, but is not necessary to be understood for the present discussion. The upshot is that the laser's energy supply (beginning as electrical energy) is eventually converted, via excitation of the optical cavity mode, into electromagnetic wave energy of a well defined frequency *and phase*, i.e., laser light is highly coherent (as defined in Section 2.2). Furthermore, it can be produced with a high intensity. Given our understanding of the nature of light, it is clear that a high intensity corresponds to large local magnitudes of the **E** and **B** fields, which have the potential to exert significant forces on the constituent charged particles within ordinary matter. This feature of laser light is exploited in the design of optical tweezers, ultimately used to exert forces that can test the mechanical properties of interesting complex fluids.

Chapter 3

Geometrical Optics

Alison Yao
SUPA and Department of Physics, University of Strathclyde, Glasgow G4 0NG, UK
alison.yao@strath.ac.uk

3.1 Introduction

In 1864, James Clerk Maxwell made the remarkable discovery that light could be described as an electromagnetic wave (Maxwell, 1864) and therefore was governed by the same laws as all other forms of electromagnetic radiation. In particular, as the electromagnetic wave equation is derived from Maxwell's equations, then any solution to the wave equation must also satisfy Maxwell's equations. These provide extra constraints on the solutions such as, for example, the relative directions and magnitudes of the electric and magnetic fields which constitute the light. Moreover, they determine how the light is affected by the medium of propagation and the behaviour at boundaries between different media, resulting in the phenomena of reflection and refraction.

In this chapter we will look in detail at the derivation of the wave equation in vacuum and in (non-conducting) dielectric media. We will then investigate the properties of the simplest solutions to

Microrheology with Optical Tweezers: Principles and Applications
Edited by Manlio Tassieri

ISBN 978-981-4669-18-4 (Hardcover), 978-981-4669-19-1 (eBook)
www.panstanford.com

these wave equations, monochromatic plane waves, and derive the laws of reflection and refraction at dielectric boundaries. Finally, we will consider more physically realistic solutions which have small spatial extent transverse to the propagation direction. A typical example is a Gaussian beam, familiar to anyone who uses a laser pointer. As we will show, we can characterise these by their beam width, w, and radius of curvature, R and thus describe the beams using *complex beam parameter (cbp)* theory. The Gaussian mode is just one of many possible solutions to the paraxial wave equation, the particular choice of solution being generally dependent on the geometry of the system. We introduce two of the most common orthonormal basis sets: the *Laguerre-Gaussian* (LG) modes and the *Hermite-Gaussian* (HG) modes. The LG modes are of particular use in cases of cylindrical symmetry, while the HG modes are more appropriate for Cartesian geometry and in situations where there is astigmatism.

3.2 Maxwell's Equations

The equations that we recognise today as Maxwell's equations:[a]

$$\nabla \cdot \mathbf{D} = \rho_f, \tag{3.1}$$

$$\nabla \cdot \mathbf{B} = 0, \tag{3.2}$$

$$\nabla \times \mathbf{E} = -\frac{\partial \mathbf{B}}{\partial t}, \tag{3.3}$$

$$\nabla \times \mathbf{H} = \mathbf{J}_f + \frac{\partial \mathbf{D}}{\partial t}. \tag{3.4}$$

are a set of four equations governing the behaviour of electric and magnetic fields. **E** is the electric field measured in volts per metre, Vm^{-1}, and **H** is the magnetic field measured in ampères per metre, Am^{-1}. **D** is the *electric displacement field*, which is related to the electric field via the polarisation, **P**, according to

$$\mathbf{D} = \varepsilon_0 \mathbf{E} + \mathbf{P}. \tag{3.5}$$

[a]When Maxwell's Equations first appeared in "A dynamical theory of the electromagnetic field", Philosophical Transactions of the Royal Society of London, in 1865 there were actually 20 equations (Maxwell, 1855). The compact vector version that we recognise today were formulated by Oliver Heaviside in 1884.

In linear dielectric media we can write the polarisation as $\mathbf{P} = \varepsilon_0 \chi \mathbf{E}$, where ε_0 is the permittivity of free space (8.85×10^{-12} Fm^{-1}) and χ is the electric susceptibility of the medium. This allows us to write

$$\mathbf{D} = \varepsilon_0 (1 + \chi) \mathbf{E} = \varepsilon_0 \varepsilon_r \mathbf{E}, \tag{3.6}$$

where ε_r is the relative permittivity, or dielectric constant, of the medium. Similarly, we can write the magnetic field as

$$\mathbf{H} = \frac{1}{\mu_0} \mathbf{B} - \mathbf{M}, \tag{3.7}$$

where **B** is the magnetic induction measured in Tesla, T, and **M** is the *magnetisation* of the medium. Writing this in terms of its relative permeability, μ_r, gives

$$\mathbf{M} = (\mu_r - 1) \mu_0 \mathbf{H} \tag{3.8}$$

$$\Rightarrow \mathbf{B} = \mu_0 \mu_r \mathbf{H}. \tag{3.9}$$

Maxwell's equations, together with the Lorentz force law

$$\mathbf{F} = q(\mathbf{E} + \mathbf{v} \times \mathbf{B}), \tag{3.10}$$

which gives the force on a charge q moving with velocity v in the presence of an electric field **E** and a magnetic induction **B**, form the foundation of the theory of classical electrodynamics.

At first sight it may seem strange that these equations are now known collectively as Maxwell's equations. Equations 3.1 and 3.2 are in fact Gauss's law and Gauss's law for magnetism, respectively, while Eq. 3.3 is Faraday's law and (3.4) is the Ampère-Maxwell law. Maxwell's seemingly innocuous contribution is simply the last term in Eq. 3.4 which predicts that changing electric fields induce magnetic fields in an analogous way to Faraday's law which states that changing magnetic fields induce electric fields.

Maxwell's groundbreaking intuition, however, led him to combine these equations to derive *wave equations* for the electric and magnetic fields. Moreover, these allowed him to make the theoretical prediction that, in air, the resultant electromagnetic waves propagate at the speed of light. This startling discovery is recorded in a letter from Maxwell to Faraday in 1861:

Letter to Michael Faraday, 19th October 1861

> *...From the determination by Kohlrausch and Weber of the numerical relation between the statical and magnetic effects of electricity (Kohlrausch and Weber, 1857), I have determined the elasticity of the medium in air, and assuming that it is the same with the luminiferous ether I have determined the velocity of propagation of transverse vibrations.*
>
> *The result is*
>
> *193088 miles per second*
>
> *(deduced from electrical & magnetic experiments) (Maxwell, 1862). Fizeau has determined the velocity of light*
>
> *= 193118 miles per second*
>
> *by direct experiment (Maxwell, 1862). This coincidence is not merely numerical. I worked out the formulae in the country, before seeing Webers number, which is in millimetres, and I think we have now strong reason to believe, whether my theory is a fact or not, that the luminiferous and the electromagnetic medium are one.*

This was an astounding accomplishment and is arguably one of the key results of 19th century science. Not only did Maxwell unify the forces of electricity and magnetism and make the leap between electromagnetism and light but he also predicted the speed of light based on properties of the "medium" of propagation (permittivity and permeability) at a time when many scientists believed in the existence of a luminiferous (a)ether through which the transverse vibrations (light) propagates.

3.3 From Maxwell's Equations to the Wave Equation

3.3.1 *Wave Equations in a Vacuum*

In vacuum,with no free charges ($\rho_f = 0$) or free currents ($\mathbf{J}_f = 0$), we can write $\mathbf{D} = \epsilon_0 \mathbf{E}$ and $\mathbf{B} = \mu_0 \mathbf{H}$. Substituting these into Eqs. 3.1–3.4 allows us to write them in terms of the electric and magnetic fields, **E** and **H**, alone:

$$\nabla \cdot \mathbf{E} = 0, \tag{3.11}$$

$$\nabla \cdot \mathbf{H} = 0, \tag{3.12}$$

$$\nabla \times \mathbf{E} = -\mu_0 \frac{\partial \mathbf{H}}{\partial t}, \tag{3.13}$$

$$\nabla \times \mathbf{H} = \epsilon_0 \frac{\partial \mathbf{E}}{\partial t}. \tag{3.14}$$

Written in this form it is very easy to see the symmetry between the two equations for each field:

$$\nabla \cdot \mathbf{E} = 0; \quad \nabla \times \mathbf{E} = -\mu_0 \frac{\partial \mathbf{H}}{\partial t}, \tag{3.15}$$

$$\nabla \cdot \mathbf{H} = 0; \quad \nabla \times \mathbf{H} = \epsilon_0 \frac{\partial \mathbf{E}}{\partial t}. \tag{3.16}$$

Indeed, if we make the changes $\mathbf{E} \rightarrow \mathbf{H}$ and $\mathbf{H} \rightarrow -(\epsilon_0/\mu_0)\,\mathbf{E}$ then Eq. 3.15 turn into Eq. 3.16 and vice versa.

In order to derive wave equations in **E** and **H** we can also remove **H** from Eq. 3.15 to find an equation in **E** alone and **E** from Eq. 3.16 to find an equation in **H** alone. We will demonstrate this for Eq. 3.15 and leave the analogous derivation using (3.16) as an exercise for the reader. We start by taking the curl of Maxwell III (3.13):

$$\nabla \times (\nabla \times \mathbf{E}) = \nabla \times \left(-\mu_0 \frac{\partial \mathbf{H}}{\partial t}\right).$$

Consider, first, the left hand side (lhs) of this equation. This is a standard vector identity:

$$\nabla \times \ (\nabla \times \mathbf{E}\ ) = -\nabla^2 \mathbf{E} + \nabla\ (\nabla \cdot \mathbf{E}\ ). \tag{3.17}$$

For completeness, however, we show the derivation. Consider, for convenience, just the x-component:

$$
\begin{aligned}
[\nabla \times (\nabla \times \mathbf{E})]_x &= \left[\begin{vmatrix} i & j & k \\ \frac{\partial}{\partial x} & \frac{\partial}{\partial y} & \frac{\partial}{\partial z} \\ (\nabla \times \mathbf{E})_x & (\nabla \times \mathbf{E})_y & (\nabla \times \mathbf{E})_z \end{vmatrix}\right]_x \\
&= \frac{\partial}{\partial y}(\nabla \times \mathbf{E})_z - \frac{\partial}{\partial z}(\nabla \times \mathbf{E})_y \\
&= \frac{\partial}{\partial y}\left(\frac{\partial}{\partial x}E_y - \frac{\partial}{\partial y}E_x\right) - \frac{\partial}{\partial z}\left(\frac{\partial}{\partial z}E_x - \frac{\partial}{\partial x}E_z\right) \\
&= \left(-\frac{\partial^2}{\partial y^2} - \frac{\partial^2}{\partial z^2}\right)E_x + \frac{\partial}{\partial x}\left(\frac{\partial}{\partial y}E_y + \frac{\partial}{\partial z}E_z\right) \\
&= -\left(\frac{\partial^2}{\partial x^2} + \frac{\partial^2}{\partial y^2} + \frac{\partial^2}{\partial z^2}\right)E_x \\
&\quad + \frac{\partial}{\partial x}\left(\frac{\partial}{\partial x}E_x + \frac{\partial}{\partial y}E_y + \frac{\partial}{\partial z}E_z\right) \\
&= -\left[\nabla^2 \mathbf{E}\right]_x + [\nabla\ (\nabla \cdot \mathbf{E}\ )]_x.
\end{aligned} \tag{3.18}
$$

A similar result holds for the y- and z-components:

$$
[\nabla \times\ (\nabla \times \mathbf{E})]_y = -\left[\nabla^2 E\right]_y + [\nabla\ (\nabla \cdot \mathbf{E}\ )]_y
$$

$$
[\nabla \times\ (\nabla \times \mathbf{E})]_z = -\left[\nabla^2 E\right]_z + [\nabla\ (\nabla \cdot \mathbf{E}\ )]_z
$$

Finally, recombining these x-, y- and z-components we can write the lhs as

$$
\nabla \times\ (\nabla \times \mathbf{E}\ ) = -\nabla^2 \mathbf{E} + \nabla\ (\nabla \cdot \mathbf{E}\ ) \tag{3.19}
$$

exactly as in (3.17).

Since in this case we are considering Maxwell's equations with no free charges, $\rho_f = 0$, we know that $\nabla \cdot \mathbf{E} = 0$ by virtue of Maxwell I (3.11). Thus we can write (3.19) as

$$
\nabla \times\ (\nabla \times \mathbf{E}\ ) = -\nabla^2 \mathbf{E}. \tag{3.20}
$$

Now consider the right hand side of the curl of Maxwell III:

$$
\nabla \times \left(-\mu_0 \frac{\partial \mathbf{H}}{\partial t}\right) = -\mu_0 \nabla \times \frac{\partial \mathbf{H}}{\partial t} = -\mu_0 \frac{\partial}{\partial t}(\nabla \times \mathbf{H}).
$$

From Maxwell IV (3.14) we have that $\nabla \times \mathbf{H} = \varepsilon_0 \partial \mathbf{E}/\partial t$ and so we can write

$$
-\mu_0 \frac{\partial}{\partial t}(\nabla \times \mathbf{H}) = -\mu_0 \frac{\partial}{\partial t}\left(\varepsilon_0 \frac{\partial \mathbf{E}}{\partial t}\right) = -\mu_0 \varepsilon_0 \frac{\partial^2 \mathbf{E}}{\partial t^2}. \tag{3.21}
$$

Finally, equating the two sides (3.20) and (3.21), we have

$$\nabla^2 \mathbf{E} = \mu_0 \varepsilon_0 \frac{\partial^2 \mathbf{E}}{\partial t^2}. \tag{3.22}$$

If we set

$$\mu_0 \varepsilon_0 = \frac{1}{c^2}, \tag{3.23}$$

we can write this in the more familiar form

$$\nabla^2 \mathbf{E} - \frac{1}{c^2} \frac{\partial^2 \mathbf{E}}{\partial t^2} = 0. \tag{3.24}$$

Clearly then, (3.22) is a wave equation (in 3 spatial dimensions) for the electric field **E**, propagating through vacuum with speed $c = 1/\sqrt{\mu_0 \varepsilon_0}$.

Note the somewhat astounding fact that the parameters ε_0 and μ_0 were known from Coulomb's law and the Biot-Savart law, respectively–experiments on charges and currents with absolutely nothing to do with light. Using these measurements,

$$\varepsilon_0 = 8.854 \times 10^{-12}\,\mathrm{F m^{-1}}, \tag{3.25}$$

$$\mu_0 = 4\pi \times 10^{-7}\,\mathrm{N\,A^{-2}}. \tag{3.26}$$

allowed Maxwell to predict that the speed of propagation of the fields was 193088 miles per second (3×10^8 $\mathrm{ms^{-1}}$), astonishingly close to Fizeau's measurement of the speed of light (193118 miles per second) (Maxwell, 1862). Although this comes as no surprise to us, it was a startling discovery in Maxwell's day (1865) as it was the first suggestion that light consisted of electromagnetic waves.

As mentioned earlier, we can follow the same procedure to eliminate **E** from Eq. 3.16 and find a wave equation for the magnetic field, **H**:

$$\nabla^2 \mathbf{H} - \frac{1}{c^2} \frac{\partial^2 \mathbf{H}}{\partial t^2} = 0. \tag{3.27}$$

In following the derivation note the crucial contribution of Maxwell's correction to Ampère's law–without it it would not be possible to find the wave equation for **H**!

3.4 Solutions to the Wave Equation

3.4.1 *Plane Wave Solutions*

By using a Fourier series, it is possibly to write any wave as a linear (possibly infinite) sum of sinusoidal waves. For this reason we first consider the simplest possible solution to (3.24): monochromatic (single-frequency) plane waves. We first consider the case of waves travelling in the z-direction. As plane waves are uniform over every plane perpendicular to the direction of propagation they then have no x- or y-dependence and we can write the fields in the form:

$$\mathbf{E}(z, t) = \mathbf{E}_0 \cos\left[kz - \omega t + \varphi\right], \tag{3.28}$$

$$\mathbf{H}(z, t) = \mathbf{H}_0 \cos\left[kz - \omega t + \varphi\right], \tag{3.29}$$

where $k = 2\pi/\lambda$ is the wavenumber corresponding to wavelength λ, $\omega = 2\pi f$ is the angular frequency and $0 \leqslant \varphi < 2\pi$ is a phase constant. At fixed time t, the field strength varies sinusoidally with wavelength $\lambda = 2\pi/k$. At fixed position z the field strength varies sinusoidally with angular frequency ω. Note that in this case both k and ω are real. More generally (for other wave equations) k and/or ω can be complex to allow for damping or growth.

Using Euler's formula, $e^{i\theta} = \cos\theta + i\sin\theta$, we can write the fields in the mathematically more convenient form:

$$\mathbf{E}(z, t) = \mathbf{E}_0 e^{i(kz - \omega t + \varphi)}, \tag{3.30}$$

$$\mathbf{H}(z, t) = \mathbf{H}_0 e^{i(kz - \omega t + \varphi)}. \tag{3.31}$$

The *physical* fields are then the *real* part of these complex waves.

The relationship between the wavenumber (or wavelength) of a wave and its frequency, in vacuum, can be found by substituting (3.30) into Eq. 3.24:

$$k^2 = \frac{\omega^2}{c^2}. \tag{3.32}$$

This is known as the *dispersion relation*. Note that we obtain the same relation if we substitute (3.31) into Eq. 3.27.

In general the waves can travel in any arbitrary direction $\mathbf{k}$, where $\mathbf{k}$ is the wave *vector* in the direction of propagation with magnitude k, the wavenumber. We can then write monochromatic plane-wave solutions of the form

$$\mathbf{E}(\mathbf{r}, t) = \mathbf{E}_0 e^{i(\mathbf{k}\cdot\mathbf{r}-\omega t)}, \tag{3.33}$$

$$\mathbf{H}(\mathbf{r}, t) = \mathbf{H}_0 e^{i(\mathbf{k}\cdot\mathbf{r}-\omega t)}, \tag{3.34}$$

where $\mathbf{E}_0(\mathbf{r}) = (E_{0x}, E_{0y}, E_{0z})$, $\mathbf{H}_0(\mathbf{r}) = (H_{0x}, H_{0y}, H_{0z})$ are the constant vector amplitudes of the electric and magnetic fields, respectively, and the wave vector is $\mathbf{k} = (k_x, k_y, k_z)$. Note that for any point $\mathbf{r} = (x, y, z)$, $\mathbf{k}\cdot\mathbf{r} = k_x x + k_y y + k_z z$. Substituting solution (3.33) into (3.24) we recover the dispersion relation

$$k^2 = k_x^2 + k_y^2 + k_z^2 = \frac{\omega^2}{c^2}, \tag{3.35}$$

i.e. the wave vector $\mathbf{k}$ has magnitude $|\mathbf{k}| = \omega/c$.

3.4.2 *Properties of Plane Wave Solutions*

As well as the wave equations, these solutions, (3.33) and (3.34), must also satisfy the free-space Maxwell Eqs. (3.11–3.14) from which the wave equations were derived, and these impose extra constraints on the solutions.

Substituting (3.33) into Maxwell I gives

$$\begin{aligned}
&\nabla\cdot\left(\mathbf{E}_0 e^{i(\mathbf{k}\cdot\mathbf{r}-\omega t)}\right) = 0 \\
&\Rightarrow E_{0x}\frac{\partial}{\partial x}e^{i(k\cdot r-\omega t)} + E_{0y}\frac{\partial}{\partial y}e^{i(k\cdot r-\omega t)} + E_{0z}\frac{\partial}{\partial z}e^{i(k\cdot r-\omega t)} = 0 \\
&\Rightarrow i\left(E_{0x}k_x + E_{0y}k_y + E_{0z}k_z\right)e^{i(\mathbf{k}\cdot\mathbf{r}-\omega t)} = 0 \\
&\Rightarrow i\mathbf{k}\cdot\mathbf{E}e^{i(\mathbf{k}\cdot\mathbf{r}-\omega t)} = 0
\end{aligned}$$

Thus to satisfy Maxwell I the electric field $\mathbf{E}$ must be perpendicular to the wave vector $\mathbf{k}$. Similarly, Maxwell II requires that $\nabla\cdot\mathbf{H} = 0$ which gives $\mathbf{k}\cdot\mathbf{H} = 0$ so that the magnetic field $\mathbf{H}$ must also be perpendicular to the wave vector $\mathbf{k}$. Electromagnetic waves are therefore *transverse*: both the electric and magnetic fields are perpendicular to the direction of propagation.

The relationship between **E** and **H** is given by Maxwell III (& Maxwell IV):

$$\begin{aligned}\nabla \times \mathbf{E} &= -\mu_0 \frac{\partial \mathbf{H}}{\partial t} \\ \Rightarrow i\mathbf{k} \times \mathbf{E}_0 e^{i(\mathbf{k}\cdot\mathbf{r}-\omega)} &= i\omega\mu_0 \mathbf{H}_0 e^{i(\mathbf{k}\cdot\mathbf{r}-\omega t)} \\ \Rightarrow \mathbf{k} \times \mathbf{E}_0 &= \omega\mu_0 \mathbf{H}_0 \\ \Rightarrow \mathbf{H}_0 &= \frac{\mathbf{k} \times \mathbf{E}_0}{\omega\mu_0}.\end{aligned}$$

Thus the magnetic field **H** is perpendicular to both the electric field **E** and the direction of propagation **k**. The same conclusion also follows from Maxwell IV. Note also that the vector $\mathbf{E}_0 \times \mathbf{H}_0$ points in the direction of propagation, **k**.

The *magnitudes* of the free-space electric and magnetic fields are also related. From Maxwell III (see above), $\mathbf{k} \times \mathbf{E}_0 = \omega\mu_0\mathbf{H}_0$. As we have just established, **k** and $\mathbf{E}_0$ are perpendicular to each other and so we can write

$$\begin{aligned}|\mathbf{k}|\,|\mathbf{E}_0| &= \omega\mu_0\,|\mathbf{H}_0| \\ \Rightarrow \frac{|\mathbf{E}_0|}{|\mathbf{H}_0|} &= \frac{\omega\mu_0}{|\mathbf{k}|} = c\mu_0,\end{aligned}$$

using the definition of $|\mathbf{k}|$ from (3.35). As we know that $c = 1/\sqrt{\epsilon_0\mu_0}$ we can then write

$$\frac{|\mathbf{E}_0|}{|\mathbf{H}_0|} = \sqrt{\frac{\mu_0}{\varepsilon_0}} = Z_0. \tag{3.36}$$

The parameter Z_0 is known as the *impedance of free space* and has a value of $\sim 377\Omega$. Note that, using $\mathbf{B} = \mu_0\mathbf{H}$, we can also show that $|\mathbf{E}| = c|\mathbf{B}|$.

3.4.3 *Polarization*

Maxwell I and II show that the electric and magnetic fields are perpendicular to the direction of propagation, **k**. Maxwell II and IV show that the magnetic field **H** is perpendicular to both the electric field **E** and the direction of propagation **k**. We can therefore write $\mathbf{E} = E\,\hat{\mathbf{n}}$ and $\mathbf{H} = H\,(\mathbf{k} \times \hat{\mathbf{n}})$ where the unit vector $\hat{\mathbf{n}}$, known as the *polarization* vector, defines the plane of vibration of the electric field. From Maxwell I we know that **E** is transverse, i.e. perpendicular

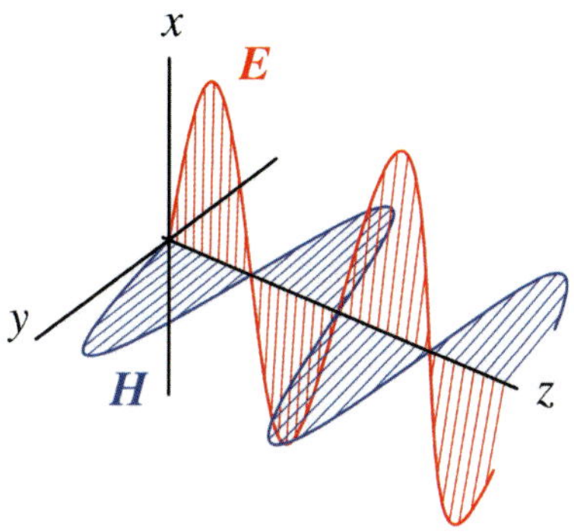

Figure 3.1 Vertical polarization.

to the wave vector **k**, so $\hat{\mathbf{n}} \cdot \mathbf{k} = 0$. Since for any given propagation direction there are two perpendicular directions, transverse waves occur in two states of polarization. This is most clearly visualised by considering waves propagating in the z-direction, as shown in Fig. 3.1, and writing $\hat{\mathbf{n}}$ in terms of the polarization angle, θ between $\hat{\mathbf{n}}$ and the x-axis:

$$\hat{\mathbf{n}} = \cos(\theta)\hat{\mathbf{x}} + \sin(\theta)\hat{\mathbf{y}}. \tag{3.37}$$

The field can then be considered as a superposition of two waves-one horizontally polarized and the other vertically. If $\theta = 0°$ the electric field is parallel to the x-axis and the wave is said to be vertically polarized. If $\theta = 90°$ the electric field is parallel to the y-axis and the wave is said to be horizontally polarized. If there is an *equal* superposition of horizontal and vertical polarizations the wave is circularly polarized as shown in Fig. 3.2.

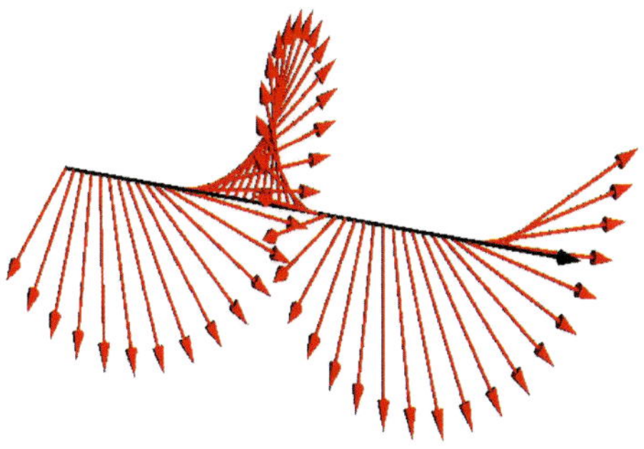

Figure 3.2 Circular polarization.

3.4.4 *Wave Equations in a Dielectric (Non-Conducting) Medium*

Another outcome of Maxwell's electromagnetic wave equation is that the speed of light depends on the properties of the medium in which it is travelling. As we saw earlier, in isotropic, homogeneous, *linear* dielectric media the electric and magnetic fields are related to the displacement find and the magnetic induction via the relative permittivity and permeability of the medium, respectively:

$$\mathbf{D} = \varepsilon_0 \varepsilon_r \mathbf{E},$$

$$\mathbf{B} = \mu_0 \mu_r \mathbf{H}.$$

With these substitutions, and again assuming no *free* charge or *free* current ($\rho_f = 0; \mathbf{J}_f = 0$), Maxwell's equations in **E** and **H** can be written as

$$\nabla \cdot \mathbf{E} = 0, \tag{3.38}$$

$$\nabla \cdot \mathbf{H} = 0, \tag{3.39}$$

$$\nabla \times \mathbf{E} = -\mu_0 \mu_r \frac{\partial \mathbf{H}}{\partial t}, \tag{3.40}$$

$$\nabla \times \mathbf{H} = \varepsilon_0 \varepsilon_r \frac{\partial \mathbf{E}}{\partial t}. \tag{3.41}$$

As before, we can find a wave equation for **E** by taking the curl of Maxwell III (3.40) and substituting in Maxwell I (3.38) and IV (3.41), with the result

$$\nabla^2 E - \varepsilon_0 \mu_0 \varepsilon_r \mu_r \frac{\partial^2 \mathbf{E}}{\partial t^2} = 0$$

$$\Rightarrow \nabla^2 \mathbf{E} - \frac{\varepsilon_r \mu_r}{c^2} \frac{\partial^2 \mathbf{E}}{\partial t^2} = 0, \tag{3.42}$$

where we have used $c = 1/\sqrt{\varepsilon_0 \mu_0}$.

Clearly this is just the wave equation again,

$$\nabla^2 \mathbf{E} - \frac{1}{v^2}\frac{\partial^2 \mathbf{E}}{\partial t^2} = 0, \tag{3.43}$$

but now with the modified velocity

$$v = \frac{c}{\sqrt{\varepsilon_r \mu_r}} = \frac{c}{n}, \tag{3.44}$$

where $n = \sqrt{\varepsilon_r \mu_r}$ is the refractive index of the medium. For most materials $n > 1$ and so the wave travels more slowly through matter. Again, there is an analogous wave equation for the magnetic field:

$$\nabla^2 \mathbf{H} - \frac{1}{v^2}\frac{\partial^2 \mathbf{H}}{\partial t^2} = 0. \tag{3.45}$$

Substituting our plane wave solution (3.33) into (3.43) we again find a dispersion relation

$$k^2 = k_x^2 + k_y^2 + k_z^2 = \frac{\omega^2}{v^2} = \frac{\omega^2 n^2}{c^2}. \tag{3.46}$$

The magnitude of the wave vector in a medium is then $|\mathbf{k}| = \omega n/c$.

Note that in addition to changing the velocity at which the wave propagates, the relative permittivity and permeability of the medium, ε_r and μ_r, also affect the characteristic impedance and hence the relative magnitudes of the electric and magnetic fields:

$$\frac{|\mathbf{E}_0|}{|\mathbf{H}_0|} = \sqrt{\frac{\mu_0 \mu_r}{\varepsilon_0 \varepsilon_r}} = \sqrt{\frac{\mu_r}{\varepsilon_r}} Z_0. \tag{3.47}$$

As ε_r is (usually) greater than 1 and in non-magnetic media $\mu_r = 1$, the impedance in the medium is generally less than that in vacuum.

3.5 Reflection and Transmission at an Interface

In Section 3.4.4 we saw that in dielectric media waves propagate with the modified velocity $v = c/n$ where the reactive index n is dependent on the type of material. In general $n > 1$ and so waves will propagate more slowly through a medium than through air. The obvious question is then what happens at the interface between air and a dielectric medium? Or, more generally, what happens at the interface between two media with different electromagnetic properties? As we might expect, we will see that electromagnetic waves can be *reflected* and *transmitted* at these interfaces.

In order to answer this question, however, we first need to know the *boundary conditions* on the fields. These can be calculated by applying Maxwell's equations in integral form:

$$\oiint_S \mathbf{D} \cdot d\mathbf{a} = Q_{\text{enc}}, \tag{3.48}$$

$$\oiint_S \mathbf{B} \cdot d\mathbf{a} = 0, \tag{3.49}$$

$$\oint_C \mathbf{E} \cdot dl = -\frac{d}{dt} \iint_S \mathbf{B} \cdot d\mathbf{a}, \tag{3.50}$$

$$\oint_C \mathbf{H} \cdot dl = I_{\text{enc}} + \frac{d}{dt} \iint_S \mathbf{D} \cdot d\mathbf{a}. \tag{3.51}$$

For the first pair S is any closed surface, while for the second pair C is the closed loop bounding the surface, S. We restrict our analysis to uncharged, non-conducting media so that there are *no surface charges or currents*. In this case the boundary conditions can be shown to be:

$$D_1^{\perp} = D_2^{\perp} \; ; \quad \mathbf{E}_1^{||} = \mathbf{E}_2^{||}, \tag{3.52}$$

$$B_1^{\perp} = B_2^{\perp} \; ; \quad \mathbf{H}_1^{||} = \mathbf{H}_2^{||}. \tag{3.53}$$

that is, the normal components of **D** and **B** and the tangential components of **E** and **H** must be continuous across the interface. It is beyond the scope of this book to derive these, but we encourage the reader to consult any good book on electromagnetism such as (Griffiths, 1998; Jackson, 1998).

We can derive the laws governing reflection and refraction by applying these boundary conditions at the interface between two linear media. Note that for isotropic, homogeneous dielectric media we can use the relations between the fields in dielectric media, $\mathbf{D} = \varepsilon_0 \varepsilon_r \mathbf{E} = \varepsilon \mathbf{E}$ and $\mathbf{B} = \mu_0 \mu_r \mathbf{H} = \mu \mathbf{H}$, to write these in terms of the electric and magnetic fields alone:

$$\varepsilon_1 E_1^{\perp} = \varepsilon_2 E_2^{\perp} \; ; \; \mathbf{E}_1^{||} = \mathbf{E}_2^{||}, \tag{3.54}$$

$$B_1^{\perp} = B_2^{\perp} \; ; \; \frac{1}{\mu_1} \mathbf{B}_1^{||} = \frac{1}{\mu_1} \mathbf{B}_2^{||}. \tag{3.55}$$

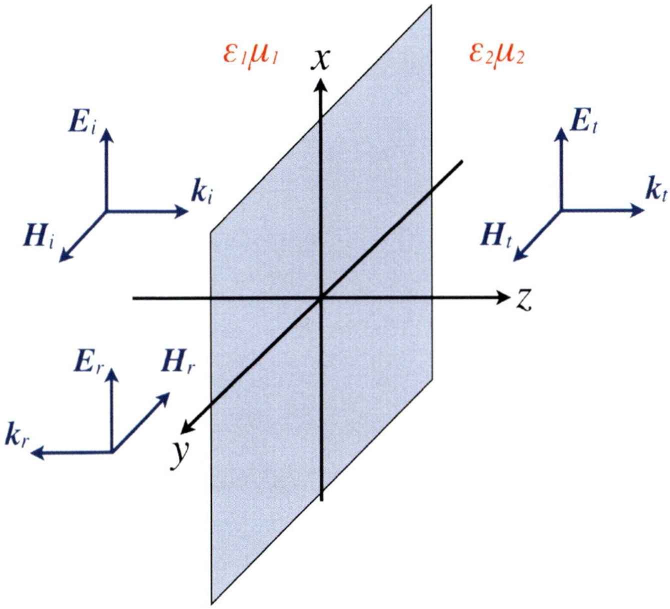

Figure 3.3 Normal incidence.

3.5.1 *Normal Incidence*

The simplest case to analyse is that of normal incidence, i.e the wave propagating in a direction **k** that is *perpendicular* to the surface so that **E** and **H** are *parallel* to the surface, as shown in Fig. 3.3. Consider a plane wave in medium 1 with frequency ω_i, travelling in the z-direction and polarised in the x-direction so that the electric and magnetic fields can be written:

$$\mathbf{E}_i = E_{0i}\hat{\mathbf{x}}e^{i(k_1 z-\omega_i t)} \,;\, \mathbf{H}_i = H_{0i}\hat{\mathbf{y}}e^{i(k_1 z-\omega_i t)} \tag{3.56}$$

This gives rise to a *reflected* wave which travels in the opposite direction in medium 1:

$$\mathbf{E}_r = E_{0r}\hat{\mathbf{x}}e^{i(-k_1 z-\omega_r t)} \,;\, \mathbf{H}_r = -H_{0r}\hat{\mathbf{y}}e^{i(-k_1 z-\omega_r t)} \tag{3.57}$$

and a *transmitted* wave which continues in the same direction in medium 2:

$$\mathbf{E}_t = E_{0t}\hat{\mathbf{x}}e^{i(k_2 z-\omega_t t)} \,;\, \mathbf{H}_t = H_{0t}\hat{\mathbf{y}}e^{i(k_2 z-\omega_t t)} \tag{3.58}$$

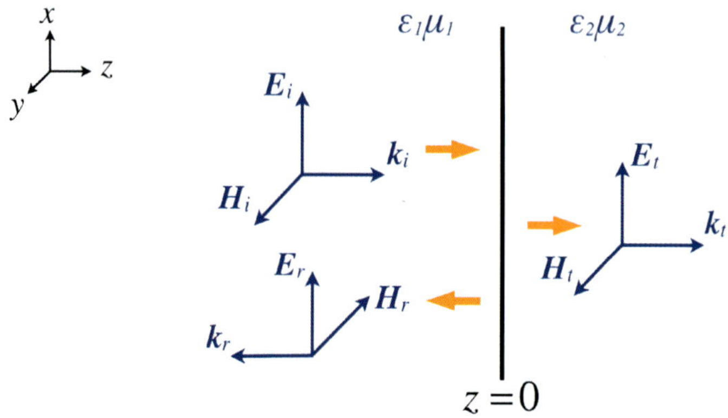

Figure 3.4 Normal incidence.

Note that the wavevector is $\mathbf{k}_1$ in medium 1 and $\mathbf{k}_2$ in medium 2. The origin of the minus sign in $\mathbf{H}_r$ is because the reflected magnetic field must point in the opposite direction in order to satisfy Maxwell's equations.

We set $z = 0$ at the plane of incidence. At this point, the combined fields on the left must match those on the right in accordance with the boundary conditions:

$$\text{(i) } D_1^\perp = D_2^\perp\text{; (ii) } B_1^\perp = B_2^\perp\text{; (iii) } \mathbf{E}_1^\parallel = \mathbf{E}_2^\parallel\text{; (iv) } \mathbf{H}_1^\parallel = \mathbf{H}_2^\parallel.$$

Since there are no components of the fields perpendicular to the surface, (i) and (ii) are trivial while (iii) and (iv) require that the tangential components of **E** and **H** are continuous. As **E** only has components in the x-direction and **H** only has components in the y-direction we can write the boundary conditions as:

$$\text{(iii) } E_{0i} + E_{0r} = E_{0t}, \tag{3.59}$$

$$\text{(iv) } H_{0i} - H_{0r} = H_{0t}. \tag{3.60}$$

As these must be true at all times, there can be no change of frequency on transmission or reflection, i.e. $\omega_i = \omega_r = \omega_t = \omega$.

In order to find a relationship between the incident field and the reflected and transmitted fields we require to express both boundary conditions in terms of the electric field. We thus use Maxwell III to express **H** in terms of **E** and rewrite boundary condition (3.60) in terms of the electric field. For a plane wave of the form $e^{i(\mathbf{k}\cdot\mathbf{r}-\omega t)}$ Maxwell III becomes

$$i\mathbf{k}\times\mathbf{E} = i\omega\mu\mathbf{H}$$
$$\Rightarrow \quad \mathbf{H} = \frac{\mathbf{k}\times\mathbf{E}}{\omega\mu}.$$

Thus we can write (3.60) as

$$\frac{k_1 E_{0i}}{\omega\mu_1} - \frac{k_1 E_{0r}}{\omega\mu_1} = \frac{k_2 E_{0t}}{\omega\mu_2}, \tag{3.61}$$

where we have made use of the fact that **E** is perpendicular to **k** at all times. From the dispersion relation (3.46) we know that $k = \omega n/c$ where $n = \sqrt{\varepsilon_r\mu_r}$ and $1/c = \sqrt{\varepsilon_0\mu_0}$. Thus we can write:

$$\begin{aligned}\frac{k}{\omega\mu} &= \frac{\omega n}{c}\frac{1}{\omega\mu}\\ &= \frac{\sqrt{\varepsilon_r\mu_r}\sqrt{\varepsilon_0\mu_0}}{\mu_0\mu_r}\\ &= \sqrt{\frac{\varepsilon}{\mu}}.\end{aligned}$$

Substituting this into Eq. 3.61 allows us to write the magnetic-field boundary condition in terms of the electric field and the properties of the medium:

$$\begin{aligned}\sqrt{\frac{\varepsilon_1}{\mu_1}}(E_{0i} - E_{0r}) &= \sqrt{\frac{\varepsilon_2}{\mu_2}}E_{0t}\\ \Rightarrow E_{0i} - E_{0r} &= \sqrt{\frac{\mu_1}{\varepsilon_1}}\sqrt{\frac{\varepsilon_2}{\mu_2}}E_{0t}\\ &= \frac{Z_1}{Z_2}E_{0t},\end{aligned}$$

where $Z_{1,2} = \sqrt{\mu_{1,2}/\varepsilon_{1,2}}$ is the impedance of medium 1,2.

Boundary conditions (iii) and (iv) have thus given us

$$E_{0i} + E_{0r} = E_{0t}, \tag{3.62}$$
$$E_{0i} - E_{0r} = \frac{Z_1}{Z_2}E_{0t}. \tag{3.63}$$

Adding these together allows us to find the transmitted field in terms of the incident field:

$$2E_{0i} = E_{0t}\left(1 + \frac{Z_1}{Z_2}\right)$$
$$\Rightarrow E_{0t} = \frac{2Z_2}{Z_2 + Z_1}E_{0i}. \tag{3.64}$$

Substituting this back in to (3.62) then gives the reflected field in terms of the incident field:

$$E_{0r} = E_{0t} - E_{0i} = \left(\frac{2Z_2}{Z_2 + Z_1} - 1 \right) E_{0i}$$

$$\Rightarrow E_{0r} = \left(\frac{Z_2 - Z_1}{Z_2 + Z_1} \right) E_{0i}. \qquad (3.65)$$

Note that for non-magnetic media ($\mu_1 = \mu_2 = \mu_0$) we can write these relations in terms of the refractive indices, $n = \sqrt{\varepsilon_r}$, of the media. The impedance of the medium, Z, is

$$Z = \sqrt{\frac{\mu}{\varepsilon}} = \sqrt{\frac{\mu_0}{\varepsilon_0 \varepsilon_r}} = \frac{1}{n}\sqrt{\frac{\mu_0}{\varepsilon_0}} = \frac{1}{n} Z_0,$$

and so we can write $Z_1/Z_2 = n_2/n_1$. The reflected and transmitted electric fields in terms of the incident field and the refractive indices of the dielectric media are then:

$$E_{0r} = \left(\frac{n_1 - n_2}{n_1 + n_2} \right) E_{0i}. \qquad (3.66)$$

$$E_{0t} = \frac{2n_1}{n_1 + n_2} E_{0i}, \qquad (3.67)$$

The fraction of incident energy that is reflected from the dielectric surface is known as the reflection coefficient, R. As the energy flux is given by the Poynting vector: $\mathbf{S} = \mathbf{E} \times \mathbf{H}$ we can calculate R from the ratio of the *cycle-average* Poynting vector for the incident and reflected fields:

$$R = \frac{\langle \mathbf{E}_r \times \mathbf{H}_r \rangle}{\langle \mathbf{E}_i \times \mathbf{H}_i \rangle} = \frac{1}{2}\frac{E_{0r}^2}{Z_1} \Big/ \frac{1}{2}\frac{E_{0i}^2}{Z_1} = \frac{E_{0r}^2}{E_{0i}^2} \qquad (3.68)$$

where we have used the definition of the impedance of free space, Z_0, (3.36) to write this in terms of E_0 alone. Substituting in our expressions for E_{0i}, E_{0r} from Eqs. 3.66 and 3.66 we find

$$R = \left(\frac{n_1 - n_2}{n_1 + n_2} \right)^2. \qquad (3.69)$$

Similarly, the ratio of transmitted to incident intensity is given by the *transmission coefficient*, T:

$$T = \frac{1}{2}\frac{E_{0t}^2}{Z_2} \Big/ \frac{1}{2}\frac{E_{0i}^2}{Z_1} = \frac{Z_1}{Z_2}\frac{E_{0t}^2}{E_{0i}^2}$$
$$\Rightarrow T = \frac{4Z_1Z_2}{(Z_2 + Z_1)^2} = \frac{4n_1n_2}{(n_1 + n_2)^2}. \quad (3.70)$$

Note that

$$R + T = \left(\frac{n_1 - n_2}{n_1 + n_2}\right)^2 + \frac{4n_1n_2}{(n_1 + n_2)^2} = 1, \quad (3.71)$$

as is required by conservation of energy.

3.5.2 *Oblique Incidence*

In practice, of course, the wave may be incident at any arbitrary angle θ_i to the boundary, of which normal incidence $\theta_i = 90°$ is just a special case. This is known as *oblique incidence* and results in reflected and transmitted waves at angles θ_r and θ_t, respectively, to the dielectric boundary, as shown in Fig. 3.5. By applying the

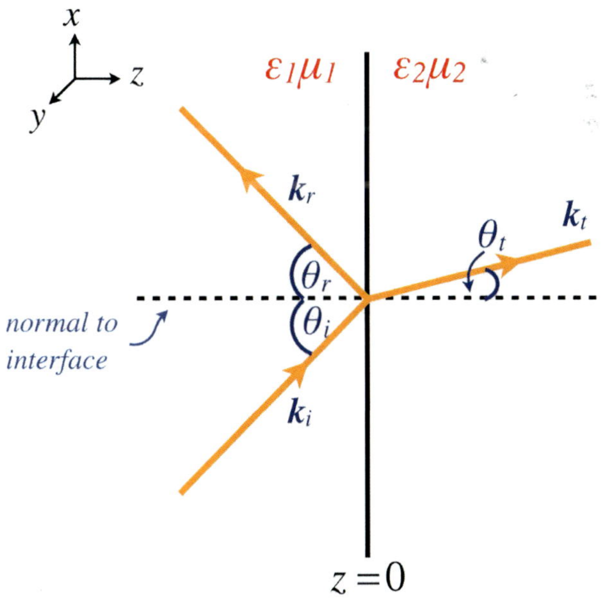

Figure 3.5 Oblique incidence.

boundary conditions in this case we can can understand and, indeed, derive the laws governing reflection and refraction.

The first of these derives from the fact that, as we said in Section 3.5.1, not only must the boundary conditions at $z = 0$ be true at all *times* so that the frequencies of the waves don't change on transmission or reflection: $\omega_i = \omega_r = \omega_t = \omega$ but they should also be true at all *positions* on the dielectric boundary. In other words, the *phases* of the waves must match:

$$(\mathbf{k}_i \cdot \mathbf{r})_{z=0} = (\mathbf{k}_r \cdot \mathbf{r})_{z=0} = (\mathbf{k}_t \cdot \mathbf{r})_{z=0}$$

$$\Rightarrow k_{ix}x + k_{iy}y = k_{rx}x + k_{ry}y = k_{tx}x + k_{ty}y. \tag{3.72}$$

This, in turn, requires that the components in each direction are equal, i.e.

$$x = 0 \quad \Rightarrow \quad k_{iy} = k_{ry} = k_{ty}, \tag{3.73}$$

$$y = 0 \quad \Rightarrow \quad k_{ix} = k_{rx} = k_{tx}. \tag{3.74}$$

As a result, all three wavevectors, $\mathbf{k}_i$, $\mathbf{k}_r$ and $\mathbf{k}_t$, must lie in the same plane. This is known as the *plane of incidence* and also includes the normal to the surface.

From the boundary conditions (3.73, 3.74) we know that in the plane of the interface, $k_{ix} = k_{rx} = k_{tx}$. Writing these in terms of the wave vectors:

$$\Rightarrow |\mathbf{k}_i| \sin\theta_i = |\mathbf{k}_r| \sin\theta_r = |\mathbf{k}_t| \sin\theta_t \tag{3.75}$$

where

θ_i is the angle of *incidence*,
θ_r is the angle of *reflection* and
θ_t is the angle of *transmission* or *refraction*

all measured w.r.t. the normal to the interface. Since for the reflected wave we have $|\mathbf{k}_i| = |\mathbf{k}_r|$ we must then have $\theta_i = \theta_r$. This results in the **law of reflection**: *the angle of incidence equals the angle of reflection.*

As $|\mathbf{k}| = \omega n/c$ and the frequency of the wave doesn't change on transmission, $\omega_i = \omega_t = \omega$, we can write the boundary condition (3.75), $|\mathbf{k}_i| \sin\theta_i = |\mathbf{k}_t| \sin\theta_t$, as

$$n_1 \sin\theta_i = n_2 \sin\theta_t.$$

This is *Snell's Law*, or the *law of refraction*:

$$\frac{\sin\theta_t}{\sin\theta_i} = \frac{n_1}{n_2}. \tag{3.76}$$

3.5.2.1 Magnitudes of the transmitted and reflected fields

The *magnitudes* of the transmitted and reflected fields depend on the *orientations* of the electric and magnetic fields. There are two cases: **E** *perpendicular* to the plane of incidence, "s" polarisation, and **E** *parallel* to the plane of incidence, "p" polarisation. We begin by choosing the dielectric interface to be in the x-y-plane and the plane of incidence to be in the x-z-plane, as shown in Fig. 3.5, and applying the boundary conditions: $D^{\perp}$ continuous, $B^{\perp}$ continuous, $\mathbf{E}^{\parallel}$ continuous, $\mathbf{H}^{\parallel}$ continuous. Note that with our particular choice that the interface is in the x-y-plane then components perpendicular to the interface are in the z-direction and components parallel to the interface are in the x-y plane.

As before, we can use Maxwell III, $\mathbf{H} = (\mathbf{k} \times \mathbf{E})/\omega\mu$, to write all of the boundary conditions in terms of the x, y, z components of the electric field, **E** alone:

$$\text{(i) } D^{\perp} \text{ continuous}: \ \varepsilon_1 (E_i + E_r)_z = \varepsilon_2 (E_t)_z \tag{3.77}$$

$$\begin{aligned}\text{(ii) } B^{\perp} \text{ continuous}: \ (\mathbf{k}_i \times \mathbf{E}_i)_z + (\mathbf{k}_r \times \mathbf{E}_r)_z \\ = (\mathbf{k}_t \times \mathbf{E}_t)_z\end{aligned} \tag{3.78}$$

$$\text{(iii) } \mathbf{E}^{\parallel} \text{ continuous}: \ (\mathbf{E}_i + \mathbf{E}_r)_{x,y} = (\mathbf{E}_t)_{x,y} \tag{3.79}$$

$$\begin{aligned}\text{(iv) } \mathbf{H}^{\parallel} \text{ continuous}: \ \frac{1}{\mu_1}\left[(\mathbf{k}_i \times \mathbf{E}_i)_{x,y} + (\mathbf{k}_r \times \mathbf{E}_r)_{x,y}\right] \\ = \frac{1}{\mu_2}(\mathbf{k}_t \times \mathbf{E}_t)_{x,y},\end{aligned} \tag{3.80}$$

where we have cancelled out the common factors in ω.

3.5.2.2 *E Perpendicular* to plane of incidence

We consider first the case where **E** is *perpendicular* to the plane of incidence: "s" polarisation Since we have chosen the axes such that the incident wave vector $\mathbf{k}_i$ lies in the x-z plane (i.e., $\mathbf{k}_i = (k_{ix}, 0, k_{iz})$) then so do $\mathbf{k}_r$ and $\mathbf{k}_t$ and we can write:

$$\begin{aligned}\mathbf{k}_i &= k_i (\sin\theta_i, 0, \cos\theta_i) \\ \mathbf{k}_r &= k_r (\sin\theta_i, 0, -\cos\theta_i) \\ \mathbf{k}_t &= k_t (\sin\theta_t, 0, \cos\theta_t)\end{aligned}$$

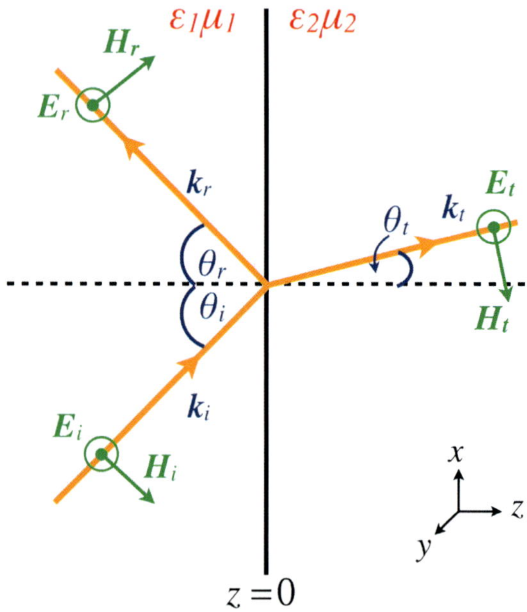

Figure 3.6 **E** *perpendicular* to the plane of incidence: "s" polarisation.

In this case, as **E** is perpendicular to the plane of incidence, and hence parallel to the interface, we have:

$$\mathbf{E}_i = E_i\hat{\mathbf{y}}\,;\ \mathbf{E}_r = E_r\hat{\mathbf{y}}\,;\ \mathbf{E}_t = E_t\hat{\mathbf{y}}.$$

We can now apply the boundary conditions.

- The first boundary condition, $D^{\perp}$ continuous, is trivial, $0 = 0$, as there is no z-component of the electric field.
- The third boundary condition, $\mathbf{E}^{\parallel}$ continuous, gives

$$\begin{aligned}(E_i + E_r)_y &= (E_t)_y \\ \Rightarrow E_i + E_r &= E_t.\end{aligned}$$

- The second, $B^{\perp}$ continuous, using Snell's law, duplicates the third.
- The fourth, $\mathbf{H}^{\parallel}$ continuous, gives:

$$\frac{1}{Z_1}(E_i - E_r)\cos\theta_i = \frac{1}{Z_2}E_t\cos\theta_t.$$

where we have used $k = \omega n/c$ and $n_{1,2} = \sqrt{\epsilon_{1,2}\mu_{1,2}}$ to write $k/(\omega\mu_{1,2}) = n_{1,2}/(c\mu_{1,2}) = 1/Z_{1,2}$.

Thus the boundary conditions have resulted in two equations:

$$E_i + E_r = E_t \tag{3.81}$$

$$E_i - E_r = \frac{Z_1}{Z_2}\frac{\cos\theta_t}{\cos\theta_i}E_t. \tag{3.82}$$

Once again we can calculate how much of the incident field is transmitted and how much is reflected by solving (3.81) and (3.82) for the transmitted and reflected fields in terms of the incident field. In order to write these purely in terms of the input parameters we first use Snell's law, $n_1 \sin\theta_i = n_2 \sin\theta_t$, to write the angle of transmittance in terms of the angle of incidence:

$$\Rightarrow \sin\theta_t = \frac{n_1}{n_2}\sin\theta_i$$

$$\Rightarrow \cos\theta_t = \sqrt{1 - (n_1/n_2)^2 \sin^2\theta_i}$$

so that (3.82) becomes

$$E_i - E_r = \frac{Z_1}{Z_2}\frac{\sqrt{1 - (n_1/n_2)^2 \sin^2\theta_i}}{\cos\theta_i}E_t. \tag{3.83}$$

We then find

$$E_r = \frac{n_1\cos\theta_i - \frac{\mu_1}{\mu_2}\sqrt{n_2^2 - n_1^2\sin^2\theta_i}}{n_1\cos\theta_i + \frac{\mu_1}{\mu_2}\sqrt{n_2^2 - n_1^2\sin^2\theta_i}}E_i \tag{3.84}$$

$$E_t = \frac{2n_1\cos\theta_i}{n_1\cos\theta_i + \frac{\mu_1}{\mu_2}\sqrt{n_2^2 - n_1^2\sin^2\theta_i}}E_i. \tag{3.85}$$

These can be written in a much more compact way if we make the substitutions

$$\alpha = \frac{\cos\theta_t}{\cos\theta_i} = \frac{\sqrt{1 - (n_1^2/n_2^2)\sin^2\theta_i}}{\cos\theta_i}$$

and

$$\beta = \frac{\mu_1}{\mu_2}\frac{n_2}{n_1}.$$

The resulting equations for the transmitted and reflected fields

$$E_r = \left(\frac{1 - \alpha\beta}{1 + \alpha\beta}\right) E_i \tag{3.86}$$

$$E_t = \left(\frac{2}{1 + \alpha\beta}\right) E_i. \tag{3.87}$$

are known as *Fresnel's equations* for the case that the polarisation is *perpendicular* to the plane of incidence. In the next section we will also derive the Fresnel's equations for the case that the polarisation is *parallel* to the plane of incidence. As we will see, that case it is possible to find an incident angle for which there is *no* reflected field. This is known as the *Brewster angle*.

3.5.2.3 *E Parallel* to plane of incidence

As in the previous case, we again choose the axes so that the wave vectors **k** lie in the x–z plane. This time, however, **E** is *parallel* to the

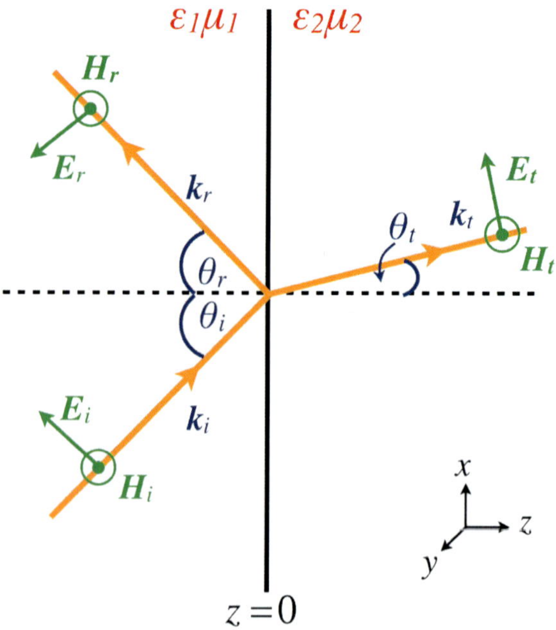

Figure 3.7 **E** *parallel* to the plane of incidence: "p" polarisation.

plane of incidence so it also lies in the x–z-plane and consequently has x- and z-components, while the magnetic field must then be *perpendicular* to the plane of incidence and thus in the y-direction.

We can again apply the boundary conditions.

- The first boundary condition, $D^{\perp}$ continuous, gives

$$\begin{aligned} \epsilon_1 \left(\mathbf{E}_i + \mathbf{E}_r\right)_z &= \epsilon_2 \left(\mathbf{E}_t\right)_z \\ \Rightarrow \epsilon_1 \left(E_i + E_r\right) \sin\theta_i &= \epsilon_2 E_t \sin\theta_t \\ \Rightarrow \sqrt{\frac{\epsilon_1}{\mu_1}} \left(E_i + E_r\right) &= \sqrt{\frac{\epsilon_2}{\mu_2}} E_t \end{aligned}$$

  using Snell's law
- The second,$B^{\perp}$ continuous, is trivial, $0 = 0$, as there is no z-component of the magnetic field.
- The third boundary condition, $\mathbf{E}^{\parallel}$ continuous, gives

$$\begin{aligned} \left(E_i + E_r\right)_x &= \left(E_t\right)_x \\ \Rightarrow \left(E_i + E_r\right) \cos\theta_i &= E_t \cos\theta_t. \end{aligned}$$

- The fourth, $\mathbf{H}^{\parallel}$ continuous, gives:

$$\begin{aligned} \left(H_i + H_r\right)_y &= \left(H_t\right)_y \\ \Rightarrow \frac{k_1}{\omega\mu_1} \left(E_i - E_r\right) &= \frac{k_2}{\omega\mu_2} E_t \\ \Rightarrow \sqrt{\frac{\varepsilon_1}{\mu_1}} \left(E_i - E_r\right) &= \sqrt{\frac{\varepsilon_2}{\mu_2}} E_t, \end{aligned}$$

  is the same as the second.

Note that we have used $H = k/(\omega\mu)E$ from Maxwell III and $k_{1,2} = \omega n_{1,2}/c$ and $n_{1,2} = \sqrt{\varepsilon_{1,2}\mu_{1,2}}$.

Thus the boundary conditions have resulted in two equations:

$$\left(E_i + E_r\right) \cos\theta_i = E_t \cos\theta_t, \tag{3.88}$$

$$\sqrt{\frac{\varepsilon_1}{\mu_1}} \left(E_i - E_r\right) = \sqrt{\frac{\varepsilon_2}{\mu_2}} E_t. \tag{3.89}$$

To find the amount of the incident field transmitted and reflected we can, as before, use these two conditions together with Snell's law to solve for the transmitted and reflected fields in terms of the incident field. We find

$$E_r = \frac{\frac{\mu_1}{\mu_2} n_2^2 \cos\theta_i - n_1\sqrt{n_2^2 - n_1^2 \sin^2\theta_i}}{\frac{\mu_1}{\mu_2} n_2^2 \cos\theta_i + n_1\sqrt{n_2^2 - n_1^2 \sin^2\theta_i}} E_i \tag{3.90}$$

$$E_t = \frac{2n_1 n_2 \cos\theta_i}{\frac{\mu_1}{\mu_2} n_2^2 \cos\theta_i + n_1\sqrt{n_2^2 - n_1^2 \sin^2\theta_i}} E_i. \tag{3.91}$$

These can again be written in a much simpler form by making the substitutions for α and β. Finally we have the Fresnel's equations for the case of polarization parallel to the plane of incidence:

$$E_r = \left(\frac{\alpha - \beta}{\alpha + \beta}\right) E_i \tag{3.92}$$

$$E_t = \left(\frac{2}{\alpha + \beta}\right) E_i. \tag{3.93}$$

3.5.2.4 Brewster's angle and total internal reflection

From Eq. 3.92, it is clear that if $\alpha = \beta$, then there will be *no* reflected wave, i.e., $E_r = 0$. It is straightforward to calculate the angle of incidence at which this will occur. In the usual case of non-magnetic media we have $\mu_1 = \mu_2 = \mu_0$. We can then write Eq. 3.90 as

$$E_r = \frac{n_2^2 \cos\theta_i - n_1\sqrt{n_2^2 - n_1^2 \sin^2\theta_i}}{n_2^2 \cos\theta_i + n_1\sqrt{n_2^2 - n_1^2 \sin^2\theta_i}} E_i.$$

Clearly, there will be *no* reflected field if:

$$n_2^2 \cos\theta_i - n_1\sqrt{n_2^2 - n_1^2 \sin^2\theta_i} = 0$$

$$\Rightarrow \sin\theta_i = \frac{n_2}{\sqrt{n_1^2 + n_2^2}}. \tag{3.94}$$

The angle at which there is no reflected wave is known as the Brewster angle, θ_B. It can also be written as

$$\Rightarrow \cos\theta_B = \frac{n_1}{\sqrt{n_1^2 + n_2^2}} \quad \text{or} \quad \tan\theta_B = \frac{n_2}{n_1}.$$

Note that this effect does not occur in the case that the electric field is polarised perpendicular to the plane of incidence. Solving Eq. 3.84 for $E_r = 0$ yields the condition that $n_1 = n_2$ in which case the media are optically indistinguishable and so there would be no reflection at any angle. This difference in behaviour between the two different polarisation states is also the mechanism behind "polarisation by reflection" as any light incident at the Brewster angle that is reflected back will be polarised perpendicular to the plane of incidence.

It is also possible to find the situation where there is no *transmitted* wave. According to Snell's law, when light passes from an optically dense medium into a less dense one ($n_1 > n_2$) then the wave vector bends *away* from the normal according to

$$\begin{aligned} n_1 \sin\theta_i &= n_2 \sin\theta_t \\ \Rightarrow \sin\theta_t &= \frac{n_1}{n_2}\sin\theta_i . \end{aligned} \quad (3.95)$$

The *critical angle* is the incident angle at which $\theta_t = 90^\circ$ such that the transmitted wave propagates exactly along the dielectric boundary. From Eq. 3.95 we can see that this corresponds to

$$\theta_c = \sin^{-1}\left(\frac{n_1}{n_2}\right).$$

Clearly, if $\frac{n_1}{n_2}\sin\theta_i > 1$ there is no solution for θ_t. This means there is no *refracted* ray, only a reflected one. This is known as *total internal reflection* (TIR). However, the fields are not zero in medium 2-there is an exponentially decaying evanescent wave.

3.6 Beam Solutions to the Wave Equation

In general, and particularly when we are dealing with lasers, we are concerned with spatially confined beams. In other words, solutions to the wave equation with finite spatial extent transverse to the propagation direction. To calculate the form of such solutions we first make the *slowly varying envelope approximation (SVEA)* that we can write the field as a product of a monochromatic carrier plane wave and envelope that varies much more slowly in space and time than the wavelength and period, respectively. Assuming, for simplicity, that the wave is propagating in the z-direction we can then write

$$\mathbf{E}(\mathbf{r}, t) = E_0(\mathbf{r})e^{i(kz-\omega t)}, \quad (3.96)$$

where $k^2 = \frac{\omega^2 n^2}{c^2}$ and the physical field is, as always, the real part. Note that most of the fast spatial variation associated with the optical field is contained in the exponential factor while the envelope, $E_0(\mathbf{r})$, represents the difference between the uniform plane wave solutions that we have examined up until now and solutions with a finite spatial extent, such as Gaussian beams.

As before, we substitute our solution into Eq. 3.43. This gives

$$\nabla_\perp^2 E_0 + \frac{\partial^2 E_0}{\partial z^2} + 2ik\frac{\partial E_0}{\partial z} - k^2 E_0 + \frac{\omega^2}{v^2} E_0 = 0, \tag{3.97}$$

where $\nabla_\perp^2 \equiv \frac{\partial^2}{\partial x^2} + \frac{\partial^2}{\partial y^2}$. Since we have made the approximation that the envelope, $E_0(\mathbf{r})$, varies very slowly in space, we can infer that:

$$\left|\frac{\partial^2 E_0}{\partial z^2}\right| \ll \left|2k\frac{\partial E_0}{\partial z}\right| \ll \left|k^2 E_0\right|,$$

thus allowing us to drop the second term in Eq. 3.97. From Eq. 3.44 we can substitute $v = c/n$, so that Eq. 3.97 can be written as:

$$\nabla_\perp^2 E_0 + 2ik\frac{\partial E_0}{\partial z} - k^2 E_0 + \frac{\omega^2 n^2}{c^2} E_0 = 0,$$

$$\Rightarrow \nabla_\perp^2 E_0 + 2ik\frac{\partial E_0}{\partial z} = 0, \tag{3.98}$$

as $k = \omega n/c$. This is known as the *paraxial wave equation*.

3.6.1 *Gaussian Beam Solutions*

The most common form of laser beam is the *Gaussian* beam, as shown in Fig. 3.8, where the transverse intensity profile of the beam takes the form:

$$I(\mathbf{r}, z) = |\mathcal{A}(\mathbf{r}, z)|^2 \exp\left(\frac{-2r^2}{w(z)^2}\right), \tag{3.99}$$

where $\mathcal{A}(\mathbf{r})$ is the amplitude on the z-axis and has dimensions of the field and $w(\mathbf{z})$ determines the width of the beam. Both change with diffraction. $r = \sqrt{x^2 + y^2}$ is the radial distance from the centre of the beam. This is the lowest order, TEM_{00}, mode, and is the most widely used in practice. Unlike other modes there are no phase reversals across the beam so it is spatially coherent. Also, its beam

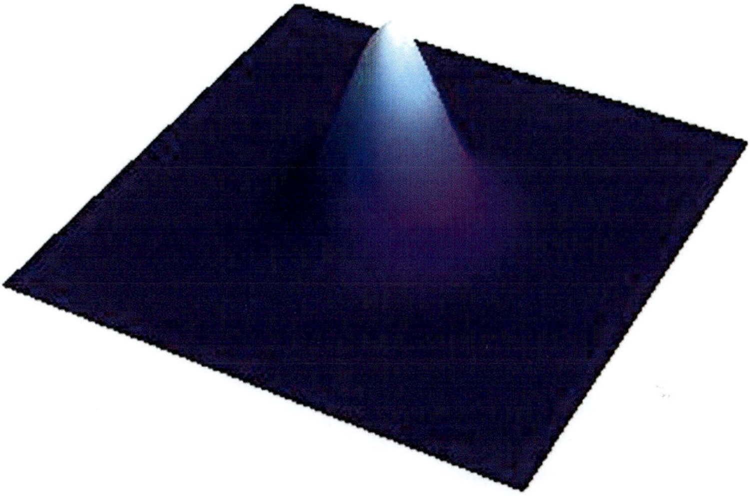

Figure 3.8 Intensity profile of a Gaussian beam.

has the smallest angular divergence and it can be focused down to the smallest spot size.

In order to find a *Gaussian* beam solution to Eq. 3.98 we assume a trial solution of the form

$$E_0(r, z) = \mathcal{A}(z) \exp\left(\frac{ikr^2}{2q(z)}\right), \tag{3.100}$$

where $\mathcal{A}(z)$ is the amplitude and $q(z)$ determines the properties of the Gaussian beam as it propagates. Note that this is not the only solution to (3.98), but it is perhaps the most important one, and is certainly the one most desired in practical laser systems. A beam with this profile is known as the *fundamental mode* of a resonator.

To obtain $\mathcal{A}(z)$ and $q(z)$ we substitute (3.100) into the paraxial wave Eq. 3.98. This gives:

$$\frac{2ik}{q} - \frac{k^2r^2}{q^2} + 2ik\left(\frac{1}{\mathcal{A}}\frac{d\mathcal{A}}{dz} - \frac{ikr^2}{2q^2}\frac{dq}{dz}\right) = 0\,.$$

Equating terms in like powers of r, we find the relations:

$$\frac{1}{\mathcal{A}}\frac{d\mathcal{A}}{dz} = -\frac{1}{q}, \tag{3.101}$$

$$\frac{dq}{dz} = 1\,. \tag{3.102}$$

Equation 3.102 has trivial solution

$$q = q_0 + z, \tag{3.103}$$

and we can then integrate (3.101) to get

$$\mathcal{A}(z) = \frac{q_0}{q_0 + z}. \tag{3.104}$$

We have thus arrived at the following expression for an electric field amplitude satisfying the paraxial wave equation:

$$E_0(r, z) = \frac{q_0}{q_0 + z} \exp\left(\frac{ikr^2}{2q(z)}\right). \tag{3.105}$$

What does this solution mean? In particular, what is the physical interpretation of $q(z)$? We expect that at very large propagation distances, z, the laser beam should be like a spherical wave from a point source. That is, it should have the form

$$E(r, z) = \frac{\exp(ikR)}{R} \exp(-i\omega t), \tag{3.106}$$

where $R = \sqrt{r^2 + z^2}$ is the radius of curvature of the phase front (see Fig. 3.9). Since the propagation distance z is very much larger than the transverse spatial size of the beam, r, we can expand the distance R as

$$R \approx z + \frac{r^2}{2z} + \ldots \tag{3.107}$$

$$\Rightarrow \frac{1}{R} \approx \frac{1}{z} - \frac{r^2}{2z^3} + \ldots \tag{3.108}$$

Since $z \gg r$ we can neglect all terms higher than quadratic in r in the phase shift factor (ikR), while in the $1/R$ denominator we can also

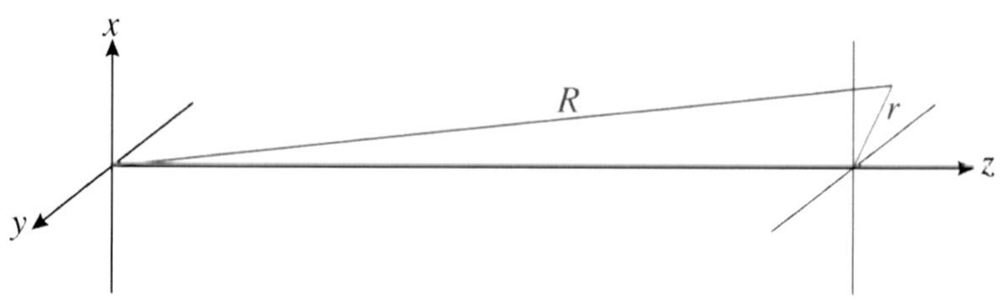

Figure 3.9 Geometry of paraxial ray propagation.

drop the term which is quadratic in r. Making these substitutions in (3.106) results in the following expression for a "paraxial-spherical" wave

$$E(r, z) \approx \frac{1}{z}\exp\left(ik\left(z+\frac{r^2}{-2z}\right)\right)\exp(-i\omega t). \qquad (3.109)$$

We now want to compare this to our solution at large z. Using the amplitude distribution from (3.105) our field will have the form

$$E(r, z) = E_0(r, z)e^{i(kz-\omega t)} = \mathcal{A}_0\frac{q_0}{q_0+z}\exp\left(\frac{ikr^2}{2q(z)}\right)e^{i(kz-\omega t)}.$$

From Eq. 3.103 we can see that in the limit that $z \rightarrow \infty$, $q \rightarrow z$ and we can write:

$$E(r, z) \approx \frac{q_0}{z}\exp\left(\frac{ikr^2}{2z}\right)\exp(ikz)\exp(-i\omega t)$$

$$= \frac{q_0}{z}\exp\left(ik\left(z+\frac{r^2}{2z}\right)\right)\exp(-i\omega t).$$

As this is clearly equivalent to Eq. 3.109, in this limit, q is the radius of curvature of the phase front near the z-axis.

From Eq. 3.105, however, it is clear that if q_0 were real then the intensity would become infinite at $z = -q_0$. In order to have a physically meaningful solution, then, we have to require that q be complex. We can thus describe the beam by:

$$E(r, z) = \mathcal{A}(z)\exp\left[ik\left(z+\frac{r^2}{2}\mathrm{Re}\left(\frac{1}{q(z)}\right)\right)\right]$$
$$\times\exp\left(\frac{-kr^2}{2}\mathrm{Im}\left(\frac{1}{q(z)}\right)\right)\exp(-i\omega t). \qquad (3.110)$$

To interpret this, first note that the second exponential term corresponds to a Gaussian with width $w(z)$. In other words, we can write

$$\exp\left(\frac{-kr^2}{2}\mathrm{Im}\left(\frac{1}{q(z)}\right)\right) = \exp\left(\frac{-r^2}{w(z)^2}\right),$$

where

$$w(z) = \left[\frac{k}{2}\mathrm{Im}\left(\frac{1}{q(z)}\right)\right]^{-1/2}$$

is called the beam width. The amplitude of the field falls to $1/e$ of its initial value at $r = w$.

To understand the first exponential term we choose the plane at $z = 0$ to be a surface of constant phase. For this to be true requires that

$$\text{Re}\left(\frac{1}{q}\right) = 0\,.$$

Thus, at a reference plane of constant phase, $z = 0$, q_0, is purely imaginary:

$$q_0 = -i\frac{kw(0)^2}{2}\;;\;\frac{1}{q_0} = \frac{2i}{kw(0)^2}\,. \tag{3.111}$$

Now as the beam propagates $q = q_0 + z$ becomes complex and hence we refer to it as the *complex beam parameter* (cbp). Away from $z = 0$, then, we have

$$\frac{1}{q} = \frac{1}{q_0 + z} = \frac{z + q_0^*}{|q_0|^2 + z^2}\,.$$

It follows that $\text{Im}(1/q)$ decreases away from $z = 0$, and thus that $w(z)$ is a *minimum* at $z = 0$. We therefore call this the *beam waist*, w_0.

From Eq. 3.110 we can see that surfaces of constant phase must obey:

$$z + \frac{r^2}{2}\text{Re}\left(\frac{1}{q(z)}\right) = \text{constant}\,.$$

Thus away from $z = 0$, $\text{Re}\left(\frac{1}{q}\right)$ must be non-zero and the phase fronts nearly parabolic, as shown in Fig. 3.10. We thus define the radius of curvature of the wavefront, $R(z)$, such that

$$\frac{1}{R(z)} = \text{Re}\left(\frac{1}{q(z)}\right)\,.$$

Finally, we can define a general expression for the complex beam parameter, q:

$$\frac{1}{q(z)} = \frac{1}{R(z)} + \frac{2i}{kw(z)^2}\,, \tag{3.112}$$

where $w(z)$ is a measure of the decrease of the field amplitude, E_0, with the distance from the axis, which is Gaussian in form. The

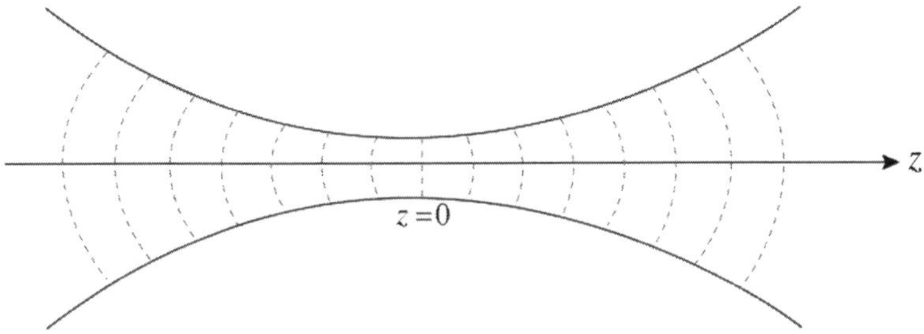

Figure 3.10 Dashed lines show the surfaces of constant phase of a Gaussian beam.

minimum diameter of the Gaussian is $2w_0$ and occurs at the beam waist where the phase front is plane.

Combining (3.103) and (3.112) and equating the real and imaginary parts allows us to write the beam width and the wavefront curvature as explicit functions of z:

$$w^2(z) = w_0^2\left[1 + \left(\frac{2z}{kw_0^2}\right)^2\right], \tag{3.113}$$

$$R(z) = z\left[1 + \left(\frac{kw_0^2}{2z}\right)^2\right]. \tag{3.114}$$

Finally we note that $\mathcal{A}(z)$ undergoes a phase change of π in propagating from large negative z through the beam waist to large positive z. The distance $z_R = kw_0^2/2$ is known as the *Rayleigh length* or *Rayleigh range*.

In order that the field reproduces itself after a roundtrip we also require that the total axial phase change is an integral multiple of 2π. To calculate this phase shift we have to re-express the amplitude of the Gaussian beam solution in terms of a real amplitude and a phase. We start by writing $\mathcal{A}(z)$ as:

$$\begin{aligned}\ln \mathcal{A}(z) &= -\ln\left(1 + \frac{z}{q_0}\right)\\ &= -\ln\left(1 + i\frac{2z}{k_0 w_0^2}\right),\end{aligned} \tag{3.115}$$

where we have used $1/q_0 = 2i/k_0 w_0^2$ from Eq. 3.111, Using the relation $\ln(a + ib) = \ln\sqrt{a^2 + b^2} + i\tan^{-1}(b/a)$ we can write

$$\ln\mathcal{A} = -\left[\ln\sqrt{1 + \left(\frac{2z}{k_0 w_0^2}\right)^2} + i\tan^{-1}\left(\frac{2z}{k_0 w_0^2}\right)\right]$$

$$\Rightarrow \mathcal{A} = \left[1 + \left(\frac{2z}{k_0 w_0^2}\right)^2\right]^{-\frac{1}{2}} \exp\left[-i\tan^{-1}\left(\frac{2z}{k_0 w_0^2}\right)\right]. \quad (3.116)$$

Finally, using the definition of $w^2(z)$ from Eq. 3.113 we can write this as

$$\mathcal{A}(z) = \frac{w_0}{w(z)}\exp(-i\eta), \quad (3.117)$$

where

$$\eta = \tan^{-1}\left(\frac{2z}{k_0 w_0^2}\right) = \tan^{-1}\left(\frac{z}{z_R}\right) \quad (3.118)$$

is a phase shift between the Gaussian beam and an ideal plane wave and $z_R = k_0 w_0^2/2$ is the *Rayleigh range*, the distance from the waist at which the beam has diffracted to twice its initial area. The amplitude factor $w_0/w(z)$ gives the expected intensity decrease on axis due to the expansion of the beam: the power in the beam is $\frac{1}{2}\pi w^2|\mathcal{A}|^2$, which we see is conserved if (3.117) holds.

We can therefore write the Gaussian envelope as:

$$E_0(r, z) = \frac{w_0}{w(z)}\exp(-i\eta)\exp\left(\frac{ik_0 r^2}{2q(z)}\right). \quad (3.119)$$

After propagation through a distance L in free space the field becomes:

$$E_0(r, L) = \frac{w_0}{w(L)}\exp\left(-i\tan^{-1}\left(\frac{L}{z_0}\right)\right)\exp\left(\frac{ik_0 r^2}{2(q_0 + L)}\right). \quad (3.120)$$

The axial phase shift is

$$\theta = k_0 L - \eta = k_0 L - \tan^{-1}\left(\frac{L}{z_0}\right). \quad (3.121)$$

The first term in this equation is just the phase shift experienced by a plane wave. The second is an additional phase shift due to the difference between a confined (Gaussian) beam and a plane wave. It goes to zero as $w_0 \to \infty$ (broad beams).

We now have an expression for a Gaussian beam solution to the paraxial wave equation. Note that the Gaussian beam is a *paraxial* approximation and that the surfaces of constant phase are paraboloids, not spheres.

3.6.2 *Higher-Order Solutions*

The fundamental Gaussian mode that we have just calculated is, however, just one of many possible solutions to the paraxial wave equation. The particular choice of solution is generally dependent on the geometry of the system. The two most common orthonormal basis sets are the *Laguerre-Gaussian* (LG) modes and the *Hermite-Gaussian* (HG) modes. The LG modes are of particular use in cases of cylindrical symmetry, while the HG modes are more appropriate for Cartesian geometry and in situations where there is astigmatism.

In two dimensions, the amplitudes of the Hermite-Gaussian modes can be written as

$$\begin{aligned} E_{mn}(x, y, z) = & \left(\frac{w_0}{w(z)}\right) H_m\left(\frac{\sqrt{2}x}{w(z)}\right) H_n\left(\frac{\sqrt{2}y}{w(z)}\right) \\ & \times \exp\left(\frac{ik_0(x^2+y^2)}{2k_0}\right) \\ & \times \exp\left(-i\left(m+n+1\right)\eta\right) . \end{aligned} \tag{3.122}$$

The last term in this equation corresponds to an additional axial phase shift,

$$\phi_{qmn} = -(m+n+1)\eta , \tag{3.123}$$

that is a function of the mode numbers. This means that the various modes of oscillation in a resonator will have different resonant frequencies. In addition to the phase shift experienced by a plane wave in the cavity, there is a difference in phase between the plane wave and a Hermite-Gaussian beam which depends on the particular Hermite-Gaussian mode. This is known as the Gouy phase (Gouy, 1890).

Figure 3.11 The lowest order Hermite-Gaussian mode, HG_{10}.

Note that the lowest order solution is

$$E_{00}(x, y, z) = \left(\frac{w_0}{w(z)}\right) \exp\left(\frac{i k_0 (x^2 + y^2)}{2k_0}\right) \exp(-i\eta),$$

which is exactly the Gaussian solution that we found earlier, (3.119).

Figure 3.12 shows the intensity plots of the first few HG modes.

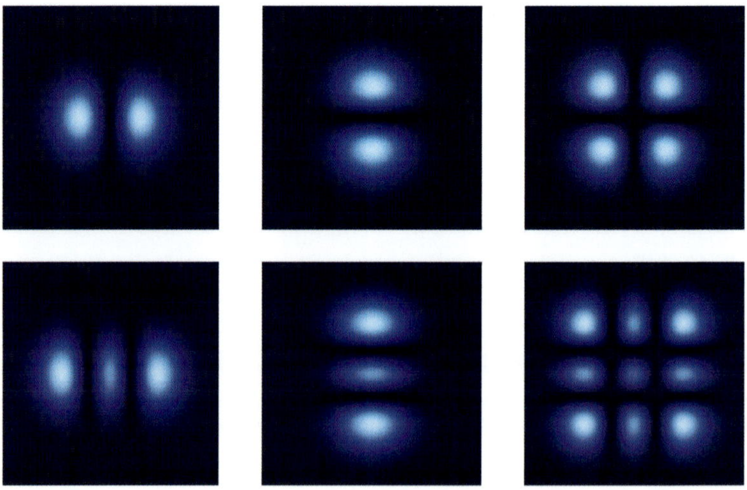

Figure 3.12 The first few Hermite-Gaussian modes: HG_{01}, HG_{10}, HG_{11} (top, left to right) HG_{02}, HG_{20}, HG_{22} (bottom, left to right).

These Hermite-Gaussian solutions form a complete and orthonormal set called the "modes of propagation" (Kogelnik and Li, 1966). Any distribution of monochromatic light may be expanded in terms of these modes by writing the field as

$$E_g(x, y, z) = \sum_m \sum_n C_{mn} E_{mn}(x, y, z), \tag{3.124}$$

where the expansion coefficients, C_{mn}, correspond to the amount of each transverse mode present in the field. Conversely, we can calculate the amount of each mode present in a particular field E_g by calculating the mode projection:

$$C_{mn} = \int_{-\infty}^{\infty} E_g(x, y, z)\, E^*_{mn}(x, y, z)\, dx\, dy. \tag{3.125}$$

The Laguerre-Gaussians (LGs) also form a complete and orthonormal set with amplitude distributions, $E_{p\ell}$, given by (Allen et al., 1999; Barnett and Zambrini, 2007; Siegman, 1990)

$$\begin{aligned} E_{p\ell} = & \sqrt{\frac{2p!}{\pi\,(p+|\ell|)!}\frac{1}{w(z)}}\left[\frac{r\sqrt{2}}{w(z)}\right]^{|\ell|} \\ & \times \exp\left[\frac{-r^2}{w^2(z)}\right] L_p^{|\ell|}\left(\frac{2r^2}{w^2(z)}\right) \exp[i\ell\phi] \\ & \times \exp\left[\frac{ik_0 r^2 z}{2(z^2+z_R^2)}\right] \\ & \times \exp\left[-i(2p+|\ell|+1)\tan^{-1}\left(\frac{z}{z_R}\right)\right] \end{aligned} \tag{3.126}$$

where $(2p + |\ell| + 1)\tan^{-1}(z/z_R)$ is the Gouy phase and $L_p^{|\ell|}(x)$ is an associated Laguerre polynomial, obtained from the more familiar Laguerre polynomials by

$$L_p^{|\ell|}(x) = (-1)^{|\ell|}\frac{d^{|\ell|}}{dx^{|\ell|}} L_{p+|\ell|}(x), \tag{3.127}$$

ℓ is the azimuthal index giving an orbital angular momentum (OAM) of $\ell\hbar$ per photon and p is the number of radial nodes in the intensity distribution.

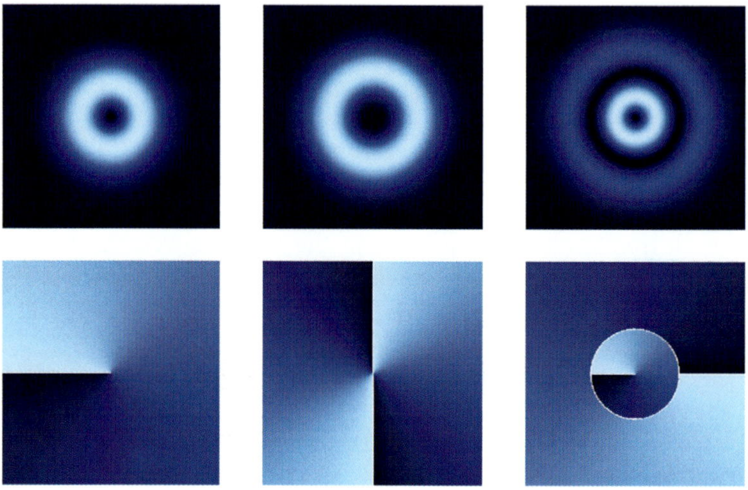

Figure 3.13 Intensity (top) and corresponding phase (bottom) plots of LG modes LG_{01}, LG_{02} and LG_{11}.

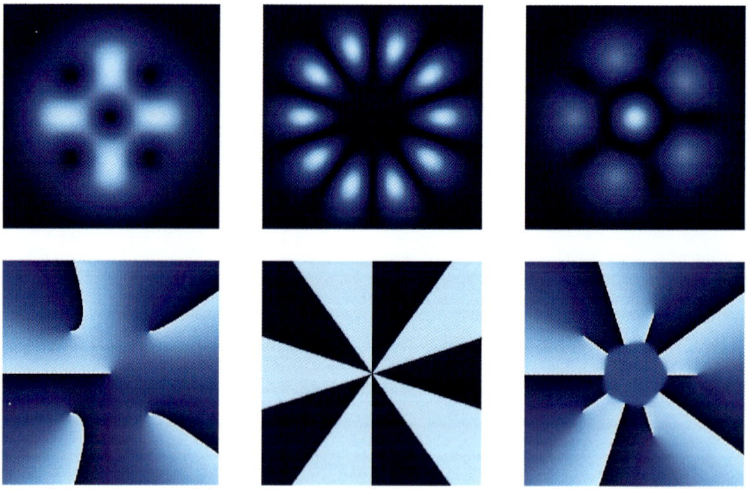

Figure 3.14 Intensity (top) and corresponding phase (bottom) plots of combinations of LG modes $LG_{01}+LG_{03}$, $LG_{0-4}+LG_{04}$, $LG_{10}+LG_{03}$ (left to right).

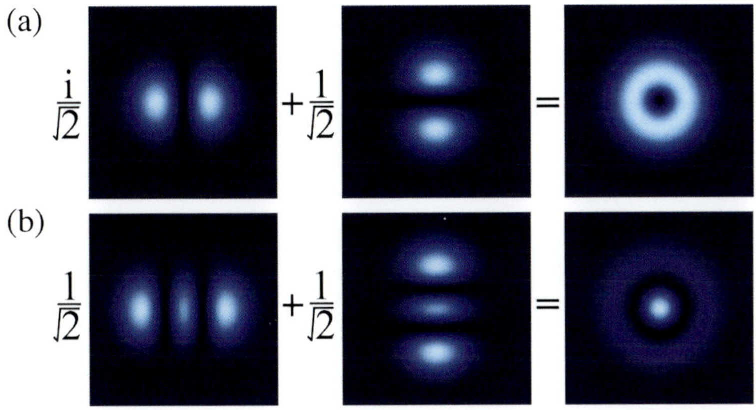

Figure 3.15 Combining Hermite-Gaussian modes to produce Laguerre-Gaussian modes. (a) $\frac{i}{\sqrt{2}}\mathrm{HG}_{01} + \frac{1}{\sqrt{2}}\mathrm{HG}_{10} = \mathrm{LG}_{01}$, (b) $\frac{1}{\sqrt{2}}\mathrm{HG}_{02} + \frac{1}{\sqrt{2}}\mathrm{HG}_{20} = \mathrm{LG}_{11}$.

In terms of their intensity cross-section, an LG mode with $\ell > 0$ comprises $p+1$ concentric rings with a zero on-axis intensity. Figure 3.13 shows the intensity and phase plots of LG modes E_{01}, E_{02} and E_{11}.

Figure 3.14 shows just a few of the interesting beam profiles that can be made by combining LG modes. Note that it is possible to transform between Hermite-Gaussian (HG) and LG modes using a specific design of cylindrical lens telescope known as a mode converter (Allen et al., 1992). This ability is based on the fact that an HG mode at 45° can be decomposed into a set of HG modes. When re-phased, this same set of HG modes can combine to form a particular LG mode. A general prescription for transforming combinations of HG modes into any particular LG mode is given in (Beijersbergen et al., 1993).

Chapter 4

Optical Forces

Michael P. Lee and David B. Phillips
School of Physics and Astronomy, SUPA, University of Glasgow, G12 8QQ, UK
mic.lee@zoho.com, david.phillips@glasgow.ac.uk

4.1 Introduction

Optical forces arise from the conservation of momentum of light scattered or absorbed by an object. If we assume any absorption is minimal, and the object is made of a transparent dielectric with a different refractive index from the surrounding medium, then the scattering encompasses light refracting through, reflecting from and diffracting around the particle. These interactions alter the direction of the incident light, and hence change its momentum, therefore, by Newton's laws of motion, a resultant force is exerted on the particle.

Each photon of light carries linear momentum, p, of $\hbar k$ kg m/s, where $\hbar$ is the reduced Plank's constant, and k is the wavenumber. Imagine a stream of photons inside a laser cavity reflecting from a mirror at one end. The photons exchange momentum with the

Microrheology with Optical Tweezers: Principles and Applications
Edited by Manlio Tassieri

ISBN 978-981-4669-18-4 (Hardcover), 978-981-4669-19-1 (eBook)
www.panstanford.com

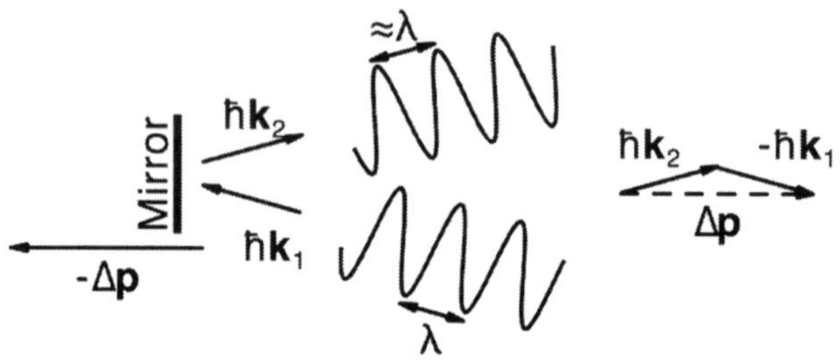

Figure 4.1 A mirror feels a small force when light elastically scatters from its surface. A force on the mirror is needed to conserve linear momentum, as the wavevector, **k**, of the photon has been changed. As the mirror has finite mass, it would be accelerated by the radiation pressure, i.e., it would gain kinetic energy. To conserve the energy (the mirror was initially at rest), the energy of the photon must be reduced, and consequently the light is redshifted. However, this lengthening of λ is typically negligible.

mirror and therefore exert a force on the cavity walls of,

$$F = 2\frac{P}{c}. \tag{4.1}$$

Here, P is the power of the light, which is related to the momentum of the individual photons via, $P = Npc$, where N is the number of photons per second and c is the speed of light. The factor of 2 is due to perfect reflection at normal incidence from the mirror inside the cavity, meaning that the photon's momentum has been changed by $2\hbar k$, see Fig. 4.1 (note that a 100% perfect reflector is technically impossible to realise in practice). When divided by the total illuminated area, this phenomenon is referred to as *radiation pressure* (Ashkin, 1970). It is interesting to note that Maxwell predicted the existence of an optical force before the concept of photons was put forward (Maxwell, 1873). In 1873, in his *Treatise on Electricity and Magnetism*, Maxwell concluded:

- In a medium in which the waves are propagated there is a pressure in the direction normal to the wave and numerically equal to the energy contained in unit volume.

Maxwell understood light to be a wave in the electromagnetic field and therefore capable of exerting a force. At this time it was already known that a water wave reflecting from a quay exerted a force on the quay, which offers a loose analogy to understand that an electromagnetic wave can push on a mirror or other scattering object.

Typically, optical forces go unnoticed as they are small compared with our everyday experience. As shown by Eq. 4.1, it takes a large number of photons to exert a force large enough to move a macro-scale object (in fact, an intensity of light not normally encountered outside of a high power laser lab). However, rather than using very high powers, the effects of optical forces from more reasonable powers can be readily witnessed on objects that are very small. A micro-particle, for example a cell, has a length of about 10^{-6} m. As this is on the order of a million times smaller than a macro-scale object, its mass is reduced by a factor of $(10^6)^3 = 10^{18}$ resulting in a proportional reduction in the magnitude of the force required to move it (Newton's second law). For small enough objects, the optical force can be sufficient to overcome gravity by a considerable margin (Ashkin, 1970; Ashkin and Dziedzic, 1971).

Arthur Ashkin was the first to show that optical microscopes have both the visualisation and beam focussing capabilities needed to realise the manipulation of objects by optical forces. Over the past three decades, an understanding of optical forces and microscopy have merged to give engineers and biologists a new tool, known as optical tweezers (or optical trapping), which relies on the high intensity gradients found in a tightly focussed laser beam to three dimensionally trap and manipulate micro-scale objects.

There are several theoretical approaches which can provide a detailed description of the light-matter interactions at work in optical tweezers (Barnett and Loudon, 2006). Depending upon the ratio of particle size to the trapping wavelength λ, different approaches can be employed. The ray-optics description of light provides an adequate description when the dimensions of the smallest features of a particle are at least $\sim$2 times larger than the trapping beam wavelength (Ashkin, 1992). For particles much smaller than λ, the optical forces felt by particles can be derived from analysis of the interaction of a particle with the electric field associated with the light (Rayleigh scattering) (Harada and Asakura, 1996). When the particle size and wavelength are of similar order, generalised Lorenz-Mie theory can be used to describe the forces on a dielectric sphere or cylinder. For particles of lower symmetry than an isotropic and homogeneous sphere, numerical simulations can be used to calculate trapping forces (Benito et al., 2008).

4.2 Gradient Forces

An important concept to help understand the way that light exerts a force in optical tweezers is that of the gradient force. Somewhat counter-intuitively, a gradient force can act "uphill", i.e. towards regions of higher intensity. To explain this, it may be instructive to first consider some other examples of the gradient force in action in macro-scale phenomena.

The so-called Coandă effect can be considered as a gradient force, as depicted in Fig. 4.2. A sphere can be held stably even in a slightly tilted upward flow of air. This is stable (i.e. when the ball moves away from equilibrium it feels a restoring force pushing it back towards equilibrium) because any rightward deviation of the ball from the centre of the jet means the jet is also deviated to the right, as shown in Fig. 4.2a. By conservation of momentum, there is a reaction force on the ball back into the flow.

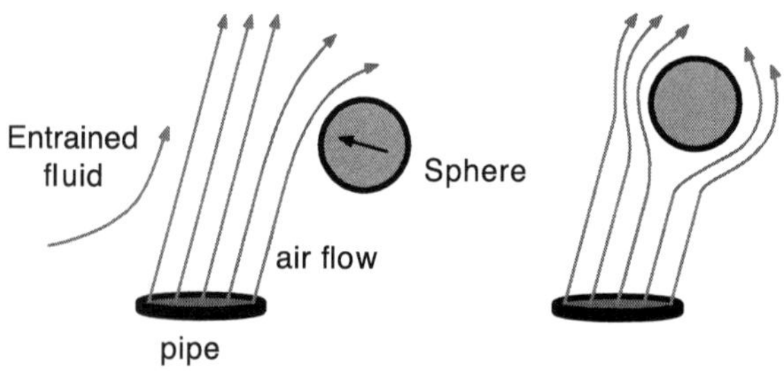

Figure 4.2 The Coandă effect. Consider the sphere when positioned out of the axis of the fastest flow as shown in (a). The presence of the sphere slows and redirects the nearby air-flow around it. Because the airflow is partly pulled towards the sphere, by conservation of momentum there is a reaction force in the opposite direction on the sphere pulling it into the flow. In the steady state case (b), the sphere is trapped laterally (i.e., orthogonally to the direction of flow) by the Coandă effect, and axially (parallel to the direction of flow) the sphere's weight is balanced by the impacts of air molecules off the sphere's surface.

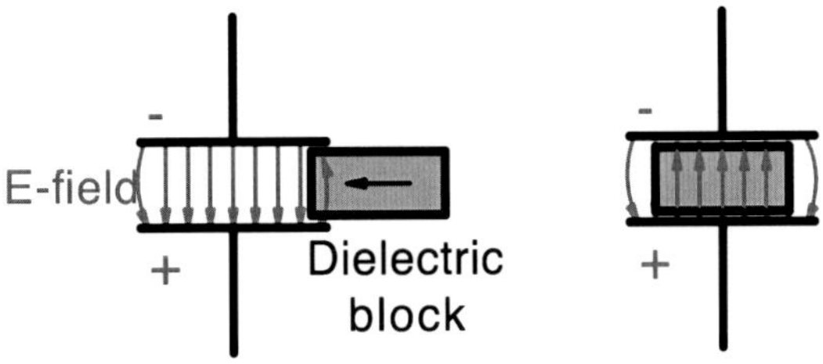

Figure 4.3 A dielectric block in an electric field. The electric field induces a dipole in the dielectric, meaning there is a force on the dielectric pulling it into the field. In the steady state case, the dielectric has moved to a position minimising the potential energy of the system.

Another example is that of a dielectric material near a static electric field tending to align itself such that the field inside the material, and hence the total energy of the system, is minimised. As shown in Fig. 4.3, the electric field polarises the material creating an internal field opposed to and balancing the external field, resulting in a net field of zero inside the material. A gradient in the external electric field across the dielectric material results in a force on it acting to pull it towards the region of highest electric field.

4.3 Ray Optics Description of Optical Tweezers

In the ray-optics model, applicable for objects that are ~ 2 or more times larger than the wavelength of the incident beam, light is considered to follow a rectilinear path, which may change direction at a boundary between two media, therefore exerting a force on the media (Ashkin, 1992; Swartzlander Jr et al., 2011). In this model, incident light is represented as a set of discrete rays. As shown in Fig. 4.4, the direction the rays travel upon striking an interface between two media of different refractive indices, n_1 and n_2, is governed by the law of reflection ($\theta_i = \theta_r$), and Snell's law of refraction:

$$\frac{\sin(\theta_i)}{\sin(\theta_t)} = \frac{n_2}{n_1}, \tag{4.2}$$

where θ_i is the angle of incidence, θ_r is the angle of reflection and θ_t is the angle of refraction, measured with respect to the line

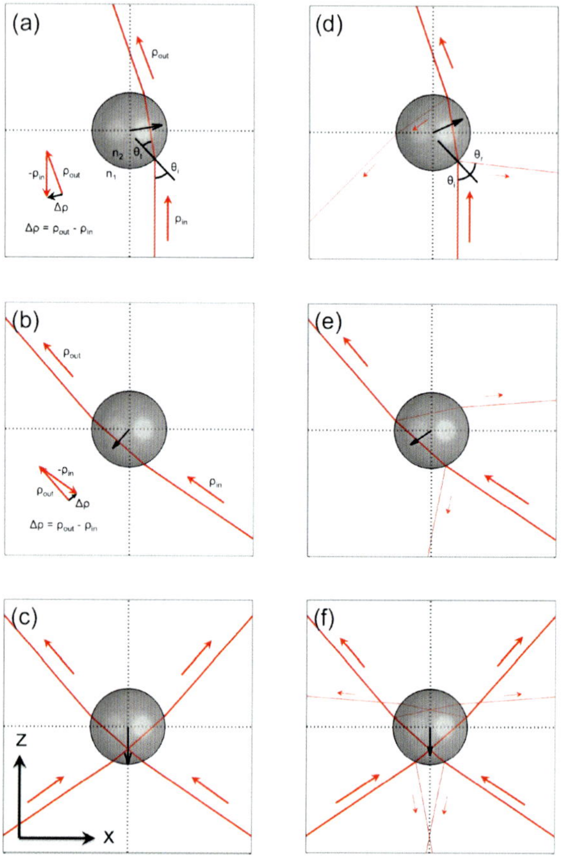

Figure 4.4 The optical reaction forces (black arrows) due to individual rays. (a) An example of a predominantly lateral reaction force on the sphere due to refraction of light through it. (b) A single incoming ray can generate both downward-axial and lateral reaction forces on a particle. (c) Two opposing rays can generate a purely axial negative reaction force on the particle. (d-f) The situations in (a-c) but now also including the first two reflections. The behaviour is qualitatively the same.

normal to the interface. In this case the light beam is traveling from medium of refractive index n1, into a medium of refractive index n2.

The fraction of light reflected at an interface is polarisation dependent and described by the Fresnel equations (here assuming a non magnetic material). R_s is the reflectance of s-polarised light, and R_p the reflectance of p-polarised light:

$$R_s = \left| \frac{n_1 \cos(\theta_i) - n_2 \cos(\theta_t)}{n_1 \cos(\theta_i) + n_2 \cos(\theta_t)} \right|^2, \tag{4.3}$$

$$R_p = \left| \frac{n_1 \cos(\theta_t) - n_2 \cos(\theta_i)}{n_1 \cos(\theta_t) + n_2 \cos(\theta_i)} \right|^2. \tag{4.4}$$

Figures 4.4 and 4.5 demonstrate how consideration of light rays changing direction at interfaces yield optical forces that result in stable optical trapping. In what follows, we refer to forces directed parallel to the optical axis of an illuminating beam as *axial* forces, shown as the vertical axis (labelled z) in Figs. 4.4 and 4.5. Similarly, we refer to forces directed orthogonally to the optical axis of an illuminating beam as *lateral* forces, and shown as the horizontal axis (labelled x in 2D plots) of Figs. 4.4 and 4.5.

Figure 4.4a-c shows how the path of individual rays can generate lateral and axial reaction forces on a spherical dielectric particle. In these diagrams all of the rays depicted lie in a plane that is parallel to the page and intersects the centre of the spherical particle. Initially, we consider only refraction of light through the particle as this dominates the behaviour in the optical trapping regime. In Fig. 4.4a, a ray incident from below is redirected as it refracts at both the incoming and outgoing interface. At the incoming interface the surface normal, and incident (θ_i) and refracted (θ_t) angles are shown. Due to the curvature of the sphere, the surface normal will be different at the entry and exit point, meaning the entrance and exit rays will not be collinear. As the momentum of the entire system is conserved, then the change in lateral momentum of the light ray results in a force with a large lateral component on the particle, illustrated by the inset vector diagram. Figure 4.4b and c demonstrate how rays can exert an axial force on a particle. In Fig. 4.4b the ray gains momentum in the axial direction after passage through the sphere, and therefore exerts a force with a negative axial component (downward) on the sphere, shown by the inset vector diagram. Figure 4.4c demonstrates how the combination of two rays incident at opposing angles from below can exert a purely axial negative force on the sphere. This situation is somewhat surprising,

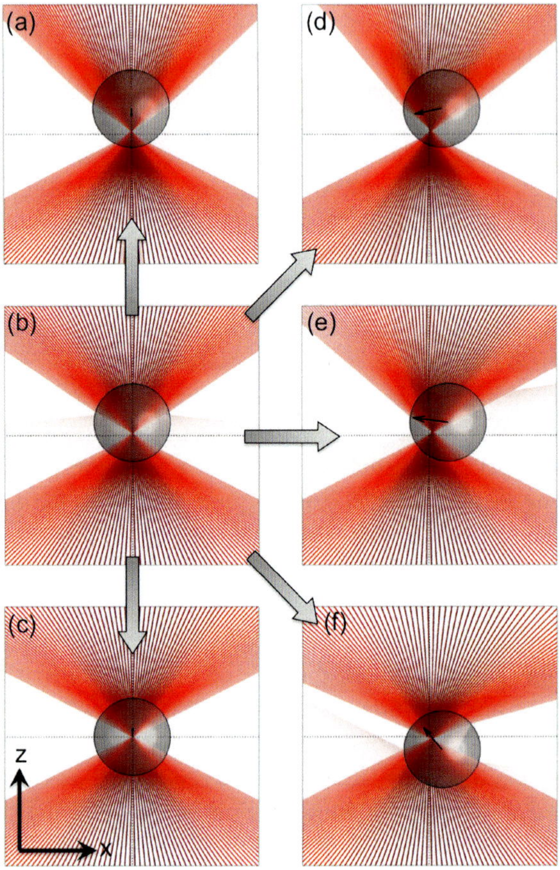

Figure 4.5 Restoring forces (black arrows) on a bead in a Gaussian focussed beam calculated using ray-optics. Each figure depicts 100 rays incident from below. (a) When the bead is displaced up, it feels a restoring force pulling it back down. (b) The position of the bead at equilibrium, slightly downstream of the focus due to radiation pressure. (c) When the centre of the bead is positioned at the focal point, the bead feels a force due to radiation pressure pushing it up. (d-f) When laterally displaced the bead also feels restoring forces pushing it back towards the vicinity of the highest intensity.

as the force on the particle is directed back towards the source of the rays.

We now consider the effect of both refraction and reflection of the rays at the interfaces. As described in the introduction, reflections can push the particle in the beam propagation direction if the beam is reflected back along the path it is incident from, via radiation pressure (Ashkin, 1970). Figures 4.4d-f show the same situations as Fig. 4.4a-c, now including contributions from the first two reflections from interfaces. The power reflected at each interface can be computed using the Fresnel Eqs. 4.3 and 4.4. The refractive index contrast between the particle (n_1) and the surrounding medium (n_2) is typically low, for example $n_{\text{water}} \approx 1.33$, and $n_{\text{silica}} \approx 1.45$. Therefore, according to Eqs. 4.3 and 4.4, the amount of power reflected is typically just a few percent at each interface. Because of this, although in reality there are more reflections than the two shown (every time a ray exits the sphere, there will also be a ray that reflects back into the sphere, refracting and reflecting at the next boundary it meets, and so on), the contribution to the momentum transfer typically becomes negligible after a small number of reflections. Figures 4.4d-f show that accounting for the first two reflections modifies the direction of the force vector, but results in essentially the same qualitative behaviour as in Figs. 4.4a-c.

The effect of a complete optical beam can be modelled by summing the contributions of individual light rays, the number of which are increased until the solution converges. As an example, Fig. 4.5 shows the optical forces exerted on a spherical particle calculated from the contribution of 100 rays, each undergoing reflections at the first two interfaces. The rays are incident from below, focussing towards the origin of each figure. As we are considering the ray-optics description, we assume that all the rays would meet at a single focal point in the absence of the sphere. We also assume perfectly monochromatic incident light, an ideal lens featuring no spherical aberrations, and no absorption of power by either medium. By weighting the contribution of each ray according to beam intensity, a model of this type can also capture the intensity profile of the beam cross-section. In this case, the rays are weighted to represent a Gaussian beam cross-section, indicated by the depth of the ray colour (darker red representing a higher intensity). This simple model is now able to capture the qualitative behaviour of an optical trapped bead. Figure 4.5 shows the sphere positioned at the focus. At this position the sphere experiences a scattering force due to radiation pressure (the effect of reflections in the model) pushing it up away from the focus. An equilibrium point is found at the position shown in Fig. 4.5, slightly down-stream of the focus, where the lateral and axial optical forces on the bead fall to zero. It is here that the radiation pressure due to reflections of the incident beam are balanced by the negative axial force shown in Figs. 4.4c and f. Figure 4.5a and d-f show the resulting optical forces as the sphere is moved in different directions away from this equilibrium position. When the bead is moved up, Fig. 4.5a shows how the spread of the transmitted ray envelope is reduced, resulting in a negative axial force on the particle pulling it back down. The reverse occurs when the particle is moved down (Fig. 4.5c). When the bead is also moved laterally as shown in Figs. 4.5d-f, we can see how the transmitted ray envelope is tilted sideways, resulting in a lateral force on the bead back towards equilibrium.

The simple model shows that the equilibrium is a stable one, as small movements of the sphere away from it result in restoring forces pushing it back towards the vicinity of the equilibrium position. The bead is always pulled back towards the vicinity of highest intensity (or slightly down-stream from that point due to radiation pressure), and so the combined action of multiple rays can be considered as a gradient force. The ray-optics model also shows that the axial restoring forces grow more slowly than the lateral restoring forces, indicating that the bead is trapped more tightly in the lateral direction than the axial one.

A more subtle point is that the trap is weakly non-conservative (Roichman et al., 2008b; Simpson and Hanna, 2007). When moved laterally, shown in Fig. 4.5 as directly right from the equilibrium position, the bead experiences a lateral restoring force, but there is also a small axial component so that the total force is not directed exactly towards the equilibrium position. Therefore the trajectory of the bead back to equilibrium, assuming no other forces are acting, is a curved path. This implies that the work required to move the bead from a lateral offset from equilibrium, along a straight path back to equilibrium, will be a subtly different from the work done when moving the bead along the curved path (remembering that work done $W = \int F(x)\delta x$, where $F(x)$ is the position dependent optical force, and x is the coordinate along the path). Therefore, if the particle travels in a closed-loop, the work done on it will not necessarily be zero, which is the signature of a non-conservative force field. However, this effect is very subtle (and indeed quite challenging to measure experimentally (Roichman et al., 2008b)), and most methods to extract useful data from trajectories of optically trapped particles assume it is moving in a conservative force-field, which is justified to within the experimental error of most experiments.

Another important point is that a negative axial gradient force, that can overcome the radiation pressure of the incoming beam, is only generated when the beam is tightly focussed and so contains incident rays that have a significant lateral component of incident momentum (see Fig. 4.5f). Therefore a *single-beam* optical trap (known as optical tweezers) requires a tightly focussed beam. In 1986, the first single-beam optical trap was created by Arthur Ashkin and co-workers (Ashkin et al., 1986). They used a tightly focussing microscope objective lens to simultaneously focus a laser beam and observe the sample, in order to optically trap and view a spherical dielectric particle.

In a weakly focussed beam, despite the small magnitude of radiation pressure, it will dominate over the axial gradient force, resulting in a bead that is laterally trapped, but also accelerated along the optical axis of the beam as shown in Figs. 4.6 and 4.7. Stable trapping can be recovered if two opposing weakly focusing Gaussian beams are used to push the particle in opposite directions.

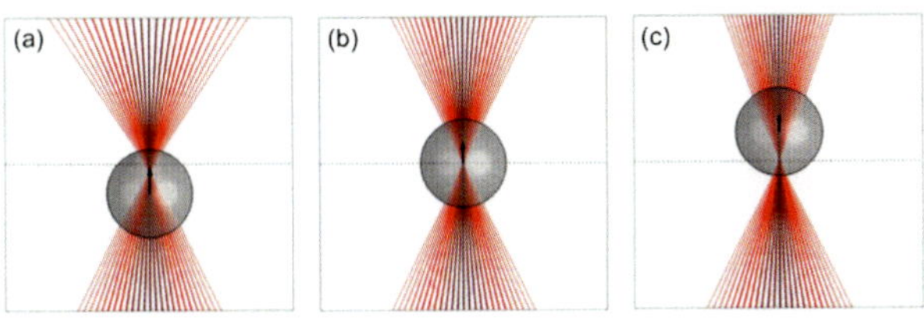

Figure 4.6 In a weakly focussed beam, the gradient forces are not great enough to overcome the radiation pressure, therefore no stable trapping point exists and the bead is propelled along the optical axis. The beam is incident from below. (a–c) shows the optical force on the bead when it is below (a), at (b) and above (c) the focal position. In all cases the optical force pushes the bead along the beam.

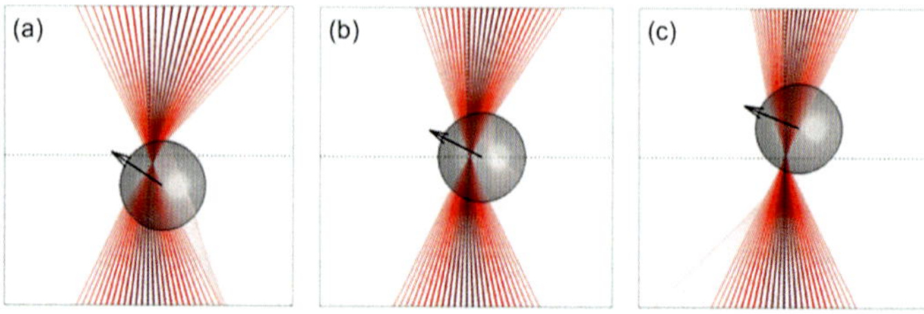

Figure 4.7 As a bead is propelled along a weakly focussed beam, lateral trapping still occurs. The beam is incident from below. (a-c) shows the optical force on the bead when it is laterally offset and below (a), at (b) and above (c) the focal position. In all cases the optical force pushes the bead along the beam, and there is also a lateral restoring force back towards the optical axis.

Here, an equilibrium position is found at the axial position where the radiation pressure from reflections of each opposing beam cancel out. Movement of the bead a small distance laterally or axially away from equilibrium result in restoring forces pushing it back to equilibrium. Such a configuration is known as a *dual-beam* optical trap, or a counter-propagating optical trap (Ashkin, 1970). This more experimentally complex, but more intuitive situation was realised by Ashkin and co-workers in 1970, sixteen years prior to their demonstration of the first single-beam optical trap. It can be created either by using two different laser beams incident from opposite directions, or a mirror to fold a second transmitted beam back on itself (Bowman et al., 2013). In order to create the optical potential well required for stable trapping, the beam waists of these counter-propagating beams need to be displaced with respect to one another, and the trap is located between the waists. This set-up allows axial positioning of the trapped particle over a greater axial range, simply by varying the relative power of the beams.

The terms "gradient force" and "scattering force" are often used to describe optical trapping in terms of forces with act to either pull the trapped object to towards the beam focus, or to push the object away. Within the ray-optics regime, it should be noted that the gradient forces do not solely arise from refraction, nor does the scattering forces solely arise from reflection. As has been seen in this section, refraction of a single ray will pull a sphere towards the ray's incident propagation axis, but also push the sphere along its direction of propagation. Reflection mainly causes additional repulsive forces, but as there are multiple reflections, there is some force pulling a particle into the focus due to these reflections (Bowman and Padgett, 2013; Lee and Padgett, 2012).

The ray-optics approach does not capture diffraction, or optical resonance of light inside a particle, and so is only appropriate when these effects are negligible. In this model, we have also assumed that no light is absorbed by the particle or surrounding mediums, which demonstrates that the origin of the optical forces are not due to the thermal heating of the particle or radiometric forces (Ashkin and Dziedzic, 1971). However, a small amount of absorption does occur, which can generate forces on objects via other mechanisms (Peterman et al., 2003). Knowledge of the optical properties of the particle and fluid guide the choice of laser wavelength to minimise these effects. For example, water has an absorption minimum at a wavelength of approximately 500 nm, and so a frequency doubled Nd:YAG laser, emitting at 532 nm is a reasonable option.

Quantification of the optical forces at work in a particular experiment can be found empirically by generalising Eq. 4.1 with a dimensionless factor, **Q**

$$\mathbf{F} = \mathbf{Q}n\frac{P}{c}, \tag{4.5}$$

where n is the refractive index of the particle. In Eq. 4.1, $|\mathbf{Q}| = 2$, while, in optical tweezers, $|\mathbf{Q}| \sim 0.1$ (Mazolli et al., 2003). **Q** depends on various factors including the size of the particle, properties of the objective, any aberrations in the optical path and the refractive index of the surrounding fluid. Calculation of **Q** within ray-optics is given by the calculation of the difference in momentum between the input an output rays. This is performed numerically and requires precise knowledge of the beam power inside the sample, which is not always easy to obtain. For this reason, it is usual to measure the optical forces experimentally (Simmons et al., 1996). Measurement of **Q** in an optical tweezers can be performed by equating the trapping force by the optical forces to the viscous drag of a surrounding fluid:

$$\mathbf{F} = \gamma\mathbf{v}, \tag{4.6}$$

where γ is the drag coefficient (for a sphere of radius r, far from any boundaries or other objects, $\gamma = 6\pi\eta r$, where η is the viscosity of the fluid) and $\mathbf{v}$ is the relative velocity at which the particle is moving through the fluid. This experiment can either be performed by moving the optical trap, or translating the sample by with automated microscope stage (see Chapter 5 for more experimental details). By translating at different speeds, it is possible to map out the force experienced by the particle as function of distance from trap centre. For the trapping configurations discussed here, for small displacements from the trap centre, the force response is approximately linear, and therefore the force can be described by,

$$\mathbf{F} = \kappa\mathbf{x}, \tag{4.7}$$

where κ is known as the trap stiffness, and typically has a value of $\sim$1 pN/nm. In the simplest, and experimentally most common situation of a single focussed Gaussian beam, the shape of the force field gives rise to a quadratic potential. However it is possible to create a variety of potential landscapes for exploring different micro-scale phenomena (Curran et al., 2012; Di Leonardo et al., 2007, 2006).

4.4 The Electric Dipole Description of Optical Tweezers

When the particle is of a similar size to the wavelength of light, λ, the ray-optics picture is no longer appropriate as diffraction effects start to play an important role in determining the optical force (Neto and Nussenzveig, 2000). The focus of a laser beam is not infinitely small (as assumed in the ray-optics model where all rays intersect at a single point at a focus), but limited by diffraction. In a focussed Gaussian beam, the diffraction limited focal spot size is quantified by the beam waist, ω_0, which defines the radius from the optical axis at which the intensity falls to $1/e^2$ of its maximum value. ω_0 is dependent upon the wavelength λ, the focal length, f, of the lens used to focus the beam, and the diameter, D, of the beam prior to focussing:

$$\omega_0 = \frac{4\lambda f}{2\pi D}, \tag{4.8}$$

For particles much smaller than wavelength of light (known as the Rayleigh regime) we can consider a single electric dipole, positioned at the centre of the particle. Such particles can be optically trapped (Dienerowitz et al., 2008; Hansen et al., 2005; Zemánek et al., 1999) As the particle is small, we assume that the electromagnetic field is constant across the particle's volume, at any given time. As described above, a dielectric particle in an electric field will become polarised (i.e., positive and negative charges are moved in opposite directions by the electric field and therefore separate), creating an internal electric field which opposes the applied field, just as is the case of a capacitor having a dielectric layer to increase the charge stored at a given voltage. For small enough particles, the charges oscillate at the same frequency as the incident light, and therefore radiate isotropically. It is this radiation that describes the light scattered by the particle, and therefore, half of the incident light will be generally scattered forwards and half will be scattered backwards, leading to an average momentum transfer of $\hbar k$ per photon being transferred from the light to the particle. This process is known as Rayleigh scattering, and it accurately describes the scattering properties of atoms and molecules, explaining, for example, why the sky is blue. In contrast to the processes of reflection and refraction in the ray-optics model, Rayleigh scattering always results in pushing the particle downstream, along the direction of propagation of the light.

The magnitude of the scattering force is proportional to the intensity of the electric field of the trapping light and the scattering cross-section of the particle,

$$\mathbf{F}_{\text{scattering}} = \frac{64\pi^5 \epsilon_0 \left|\mathbf{E}\right|^2 r^6 n_2^2}{3\lambda^4} \left(\frac{m^2 - 1}{m^2 + 2}\right)^2, \qquad (4.9)$$

where $\mathbf{E}$ is the local electric field, ϵ_0 is the permittivity of free-space, r is the particle radius, n_1 is the refractive index of the surrounding medium, n_2 is the refractive index of the particle, c is the speed of light, and $m = n_1/n_2$. $\mathbf{F}_{\text{scattering}}$ is directed perpendicular to the wavefronts of the incident light (Molloy and Padgett, 2002). The beam intensity gradient gives rise to a conceptually different force. The electric field of light at the focus of a laser beam has a high gradient. A dipole in an electric field gradient feels a force towards

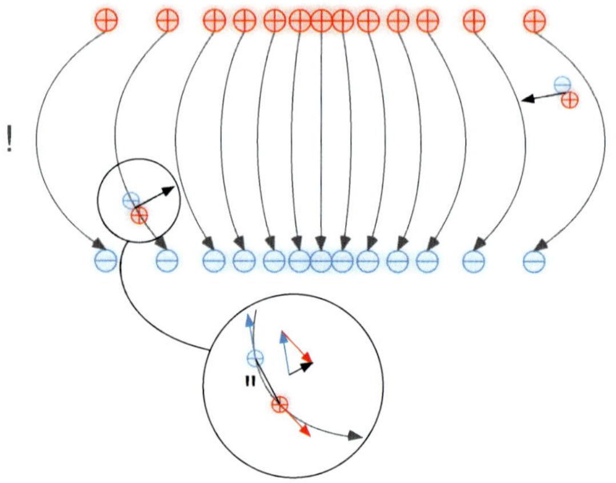

Figure 4.8 A qualitative electrostatics description of the gradient force. A dipole in an electric field is polarised. If there is an electric field gradient, in this case set up by a gradient of charges distributed in two parallel lines, then each charge in the dipole is feels an electrostatic force $\mathbf{F} = q\mathbf{E}$. When added together, this results in a force towards the region of highest electric field. A full derivation of Eq. 4.10 relies on Maxwell's equations, and also includes terms to account for the movement of charges and any time dependence of the fields.

the region of maximum intensity, Fig. 4.8. This force is given by,

$$\mathbf{F}_{\text{gradient}} = -\frac{n_2^3 r^3}{2}\left(\frac{m^2-1}{m^2-2}\right)\nabla\left(|\mathbf{E}|^2\right), \tag{4.10}$$

where $\mathbf{E}$ is the electric field. $\mathbf{F}_{\text{gradient}}$ is directed towards the region of highest intensity. As found in the ray-optics model, the equilibrium trapping point will be located on the beam axis, slightly downstream of the focal point, where $\mathbf{F}_{\text{scattering}} = \mathbf{F}_{\text{gradient}}$. To achieve stable trapping in a single focused beam, a high intensity gradient is required to balance the scattering force. This can be achieved by creating a tightly focused beam using a lens with high magnification and numerical aperture, which is equal to $n_2 \sin\theta$, where n_2 is the refractive index of the media the light is focused into, and θ is the half angle of the light cone of focusing light.

4.5 Generalized Lorenz-Mie Theory and Numerical Simulation

When the particle size is approximately equal to the wavelength of the incident light, too small for ray-optics and too large to be approximated by a single dipole, Generalised Lorenz-Mie theory can be used to calculate the forces on the particle. Generalised Lorenz-Mie theory refers to the Lorenz-Mie-Debye solution to Maxwell's equations, which involves expanding incident and scattered fields into an infinite series of spherical-vector-wave-functions and calculating series coefficients by enforcing boundary conditions on the surface of the dielectric object (Harada and Asakura, 1996). This approach is only valid for particles of high symmetry such as a sphere or cylinder. To calculate forces and torques on lower symmetry non-spherical particles (Phillips et al., 2011; Wang et al., 2013), there are many numerical simulation methods. For example: the Transfer matrix (T-matrix) technique uses a matrix to represent the relationship between incident and scattered fields, and boundary conditions are enforced to compute matrix elements (Nieminen et al., 2007; Simpson and Hanna, 2007). Finite-Difference Time-Domain (FDTD) simulations numerically solve Maxwell's equations on a lattice (Benito et al., 2008). The Discrete Dipole Approximation (DDA) (Simpson and Hanna, 2010) approximates a structure as a finite array of polarisable points. The points acquire dipole moments which are calculated from the incident electromagnetic field, and then the interactions between these dipole moments are computed to find the resulting field. In these cases, a number of simulations are performed, sweeping the object through a range of positions and orientations. The force and torque on the object in each configuration is calculated from the change in momentum between the incident and transmitted fields (Simpson, 2014). An additional method to compute optical forces is through consideration of the optical stress distributed across the surface of a dielectric particle. This can provide a simple method to estimate the trapping behaviour of arbitrarily shaped structures using the concept of the optical force density (Phillips et al., 2014).

4.6 Optical Torques

In addition to carrying $\hbar k$ units of linear momentum, photons can also carry angular momentum. This momentum has two forms, Spin Angular Momentum (SAM) and Orbital Angular Momentum (OAM) (O'Neil et al., 2002). Both of these types can be transferred to particles in an optical tweezer set up. The effect is to make the particle's spin or rotate, which is useful in mimicking biological processing, where swimming bacteria using synchronised circulating flagella (Kotar et al., 2013; Koumakis and Di Leonardo, 2013), and in measuring fluid properties, where rotational motion is less affected by nearby walls (Bishop et al., 2004; Parkin et al., 2007).

There are a number of methods which can be used to make a particle spin. For example, a birefringent particle (made from calcite or vaterite) can be rotated when illuminated with a circularly polarised laser beam (Friese et al., 1998). Circularly polarised light has a spin angular momentum of $\pm\frac{1}{2}\hbar$ per photon, i.e, the plane of the electric field rotates clockwise or anticlockwise. A birefringent particle can act in a manner similar to a quarter wave plate, converting the circularly polarised light into linear, and therefore reducing the angular momentum of the light. Therefore by conservation of angular momentum feeling a reaction torque causing it to spin. Figure 4.9 shows a schematic of this for a rectilinear beam, but it is possible to use a tightly focused beam

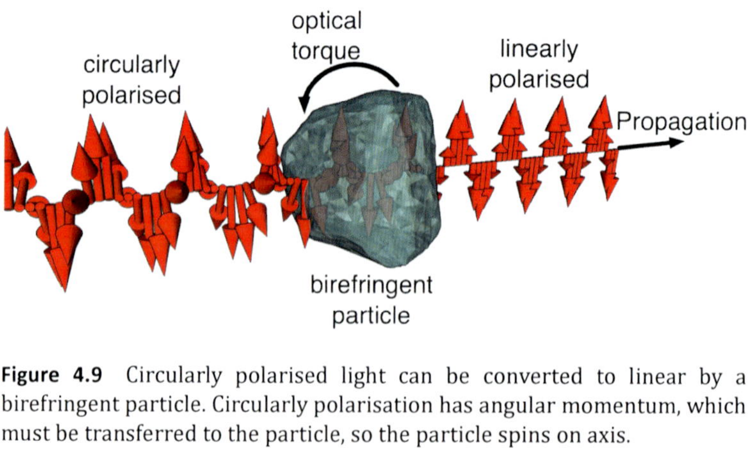

Figure 4.9 Circularly polarised light can be converted to linear by a birefringent particle. Circularly polarisation has angular momentum, which must be transferred to the particle, so the particle spins on axis.

to simultaneously trap and spin the particle on axis. In reality, the particle is not likely to convert the beam perfectly into linear polarisation (as this requires precisely the correct thickness of material the that light transmits through), but to an elliptical polarisation. However this still constitutes a change in the angular momentum of the incident beam, and therefore exerts a torque on the particle, albeit with a reduction in rotation rate of the particle from the ideal case.

Optical rotation can also be achieved by transferring some of the incident light's Orbital Angular Momentum (OAM) to translate particles on a circular path around the beam axis (Curtis and Grier, 2003; Padgett and Bowman, 2011; Paterson et al., 2001; Simpson et al., 1997). That light can carry OAM was shown in 1992 (Allen et al., 1992), and a non-zero single-mode of a Laguerre-Gaussian beam possess a twisted, helical phase front. This results in a phase singularity on-axis which causes a dark centre to the beam, known as a vortex core, surrounded by one or multiple intensity rings. The OAM of light is quantised by a topological charge, ℓ, representing the number of integer multiples of 2π the phase changes by in one circuit around the vortex core. ℓ can take any integer value. In the paraxial approximation, the twisted phase front means that the Poynting vector is slightly skewed with respect to the beam axis–that is, there is both an axial and tangential component to the Poynting vector. A particle in the beam will then feel a force around the beam axis. For highly convergent beams, it is possible to trap in 3D dimensions as well as orbit particles, although this is experimentally more difficult to achieve as the beam waist for beams carrying OAM is larger than a TEM_{00} mode. In this case, the electric field and its gradient is proportionally lower as the same power is spread over a larger area.

A third approach to achieving an optical torque is by shaping the particle (Simpson et al., 2007). This makes use of the principle that a force applied at distance away from a pivot exerts a torque. For example a chiral particle can be trapped in a linearly polarised Gaussian beam, and if aligned correctly, will spin in a manner analogous to a windmill turning in the wind. However, in the case of the particle, it is not just radiation pressure (reflection of light) causing rotation, but also refraction and diffraction too.

Therefore, the linear momentum of light rotates the micron sized chiral particle, i.e., the scattering from the particle is not isotropic but depends on the sector of the particle the light reflects and refracts from. Such particles can be fabricated using direct laser writing (Galajda and Ormos, 2001; Palima et al., 2012; Phillips et al., 2012b), and have been used, for example, to study hydrodynamic synchronisation (Di Leonardo et al., 2012).

4.7 Conclusions

In this chapter we have described how light can exert forces on matter by transfer of momentum. We have considered several theoretical approaches to predicting optical forces on dielectric particles, including ray-optics, Rayleigh scattering, Generalised Lorenz-Mie theory and numerical simulation. We also note that optical trapping of metallic nano-particles via plasmonic resonances is also an active field of research (Hansen et al., 2005; Sasaki et al., 1992; Svoboda and Block, 1994b). This can be modelled, for example using the DDA (Selhuber-Unkel et al., 2008).

In addition to optical trapping, optical forces have also found use in a variety of other situations. One simple example is the use of radiation pressure from starlight to power spacecraft with solar sails (Tsuda et al., 2011). Another example is to create an *optical spring*, which relies on optomechanical coupling to create an extremely sensitive laser cavity which can be used as a ruler to try and measure gravity waves. The mechanism behind this is as follows: Both mirrors of the cavity are freely suspended as pendulums. The mirror positions are initially held a whole number of wavelengths apart, and the cavity is "on resonance", generating the maximum build up of optical power. The set-up is arranged so that a balance is struck between the force of gravity, and the optical force due to radiation pressure on the mirror surfaces. If the cavity is lengthened, for instance by adding a static offset to the position of one mirror, the optical force on that mirror will fall and gravity will dominate, shortening the cavity. As the distance between the two mirrors shortens and the cavity once again approaches the resonance condition, radiation pressure will begin to dominate and push the mirror away. This creates a predominantly linear dependence of intra-cavity power on mirror position, which effectively rigidly couples the two mirrors together using nothing but the light fields propagating between them. This technique has been proposed to increase the quantum-limited sensitivity of large, high-power laser interferometers (Corbitt et al., 2006; Sheard et al., 2004).

In summary, despite being small enough to be largely hidden from our every day experience, optical forces are ubiquitous. They can lead to some counter-intuitive effects, such as reaction-forces on particles being directed back towards the source of the illuminating laser beam, and an understanding of their physical principles has led to the development of a range of new technologies, from non-contact manipulation of micro-scale objects, to solar sails and optical springs.

Chapter 5

Optical Tweezers Configurations

Graham M. Gibson
School of Physics and Astronomy, SUPA, University of Glasgow, G12 8QQ, UK
graham.gibson@glasgow.ac.uk

5.1 Introduction

There are many variations of optical trapping systems to suit a wide range of applications. Many systems are based on the pioneering work of Ashkin et al. (Ashkin, 1970; Ashkin and Dziedzic, 1971) but it was ever since the first realisation of a single-beam gradient-force trap, termed "Optical Tweezers" (Ashkin et al., 1986), that they have been developed as a useful tool for both the application and measurement of sensitive forces. The pico-Newton range of forces that can be applied using optical tweezers are well suited for studying biological systems (Block, 1995). They are particularly useful in microrheology where micron sized dielectric spheres are introduced into a fluid and their motion tracked using a high-speed video camera (Cicuta and Donald, 2007; Yao et al., 2009a), allowing the viscous and elastic properties of the fluid to be obtained. Microrheology techniques can be defined as either passive or active (Brau et al., 2007). Passive microrheology relies only

Microrheology with Optical Tweezers: Principles and Applications
Edited by Manlio Tassieri

ISBN 978-981-4669-18-4 (Hardcover), 978-981-4669-19-1 (eBook)
www.panstanford.com

on the thermal (Brownian) motion of the sphere whereas active microrheology relies on an external force being applied to the sphere and monitoring its response. Mechanical properties of objects can be measured through application of forces to the object itself, or to dielectric spheres attached to the object. The inherent scale of optical tweezers makes them readily applicable to microfluidic systems (Leach et al., 2006; Mushfique et al., 2008a,b) and they are suited to performing measurements on small volumes and in non conventional geometries (e.g. interior of a biological cell).

Many optical tweezers systems are built around an inverted microscope where a suitable collimated laser is focussed to a diffraction-limited spot in the image plane, providing the optical trap (see Fig. 5.1 for a typical optical tweezers configuration). The direction of the laser propagation is aligned with the microscope axis. The width of the laser beam needs to be expanded to match

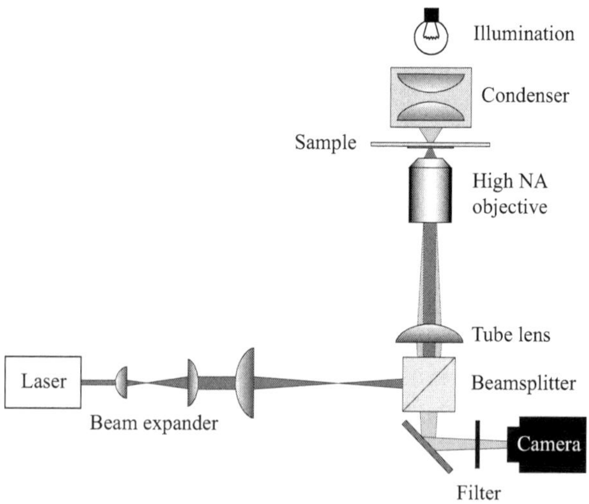

Figure 5.1 A typical optical tweezers configuration. A trapping laser beam is expanded to fill the back aperture of a high numerical aperture microscope objective, coupled into an inverted microscope using a beamsplitter or suitable dichroic mirror. A tungsten-halogen lamp and condenser is used to illuminate the sample for viewing. The motion of trapped particles in the sample is analyzed using a video camera. A filter prevents the laser from saturating the camera.

the size of the back aperture of the microscope objective lens. This is achieved using a Keplerian telescope, formed by two lenses separated by the sum of their focal lengths. The same objective lens can provide both the optical trap and an image of the object being manipulated. Many systems use a commercial light microscope with only a few simple modifications. This has the advantage of creating an optical tweezers with a high quality imaging capability and is directly compatible with many of the original imaging modes of the microscope including fluorescence, darkfield, phase contrast, differential interference contrast and epifluorescence (Kyoung et al., 2007). Additional modifications to the microscope allow the inclusion of particle position detectors such as quadrant photodiodes or high-speed video cameras, providing the capability of tracking particles and measuring the forces acting on them. A number of overviews of optical tweezers systems and design considerations can found in references (Bowman and Padgett, 2013; M-Badosa et al., 2007; Neuman and Block, 2004).

While commercial microscopes can provide a convenient solution, they are not always easy to modify. As experiments using optical tweezers become more complex, the design of the system becomes more complicated and this can require a custom microscope design. A good degree of mechanical stability is required in order to achieve nanometer scale position measurements and to detect piconewton scale forces. Many different configurations of optical tweezers are possible, dependent on the use or particular application, and include simple single trap systems (Ashkin et al., 1986, 1987), dedicated workstations based on multiple, dynamically controlled, traps (Gibson et al., 2008) to compact and very stable instruments (Gibson et al., 2012).

5.2 Different Lasers Wave-Lengths for Different Applications

The minimum requirement of an optical tweezers is coupling a laser beam into an inverted microscope. This is usually achieved using either a beamsplitter or dichroic mirror above the tube lens of the microscope to couple in the laser light while still maintaining the normal imaging capability (see Fig. 5.1). Dichroic mirrors are useful if the wavelength of the trapping laser is outside the range of wavelengths present in the illumination source, such as in the case for an IR trapping laser. A polarising beamsplitter cube can also provide a good option when using a visible wavelength trapping laser, as the laser can be efficiently coupled into the microscope without seriously attenuating the brightfield image of the sample on the camera. From the point of view of trapping objects, the most important requirement of the laser is a single mode output. A Gaussian TEM_{00} laser mode focuses to the smallest diameter beam waist, producing the most effective optical trap. When dealing with micron sized dielectric beads in water, pN range forces can be produced by $\sim$10 mW range of laser powers. There are many types of laser that can meet these requirements covering a wide range of wavelengths and output power. The choice of laser has a significant impact on the cost and performance of the system.

Compact diode-pumped solid state lasers are well suited for optical tweezers as they are efficient and can provide relatively high output powers. Their high efficiency means that less heat is transferred, via the laser housing, to the surrounding optics, minimising thermal drift. Low thermal drift of the laser output, or good pointing stability, is important in determining the limits of position or force measurements. This can be further improved by fibre coupling the laser to the microscope. Many optical tweezers systems use frequency doubled Nd:YAG or Nd:YVO_4 lasers operating at 532 nm as they can provide high powers relatively inexpensively. Diode lasers can also provide suitable powers at relatively low cost but are limited in the output mode quality. In many applications the wavelength of the trapping laser is of secondary importance. However, careful considerations are required when working with light sensitive objects like living cells, where optical damage is wavelength dependent. There can be potentially harmful heating effects due to the high light intensities in the region of the optical trap (Liu et al., 1995; Neuman and Block, 2004; Peterman et al., 2003; Schönle and Hell, 1998). Most biological materials absorb only weakly in the near infra-red making diode-pumped solid-state lasers, operating at 1064 nm, a good choice for optical tweezers

(Ashkin et al., 1987, 1990; Ferrari et al., 2005; Svoboda and Block, 1994a).

Titanium-sapphire lasers, tunable in the 700–900 nm range, have been successfully used for trapping living cells as they allow the trapping force to be optimised while minimising energy absorption (Berns et al., 1992). The lower energy absorption of the near infrared light helps to avoid unwanted cell damage that can occur with other wavelengths. Experiments have shown that live embryonic stem cells can be maintained for several hours after being manipulated using an 800 nm laser (Leach et al., 2009). However, titanium-sapphire lasers are the most expensive laser option and many currently cost in excess of £100 k.

5.3 Objectives

An essential part of a single beam optical tweezers is a high numerical aperture (NA) objective lens, used for creating a small focal volume which is tightly localised in three dimensions. Typically, 100x or 60x magnification objectives are used with an NA $\geqslant$ 1.2. Modern microscopes use objective lenses that form an image at infinity, meaning that a tube lens is required to bring the image into focus on the camera (see Fig. 5.1). The efficiency of an optical trap is limited by its axial strength. Optical traps usually require the high numerical aperture of oil-immersion objectives to provide stronger traps but these can also suffer from spherical aberrations. This can degrade the axial stiffness of the trap and affects high NA objectives more severely, making them less suited for trapping deeper into the sample (Theofanidou et al., 2004). Most microscope objectives are supplied corrected for aberrations when imaging a sample close to the microscope slide coverglass. In many applications using optical tweezers however, it is advantageous to trap deeper into the sample, where the trapped object can be kept clear of boundary surfaces. Boundary surfaces modify the drag coefficient, changing the dynamics of the trapped particle's motion as predicted by Faxén's correction (Happel and Brenner, 1983). In addition many objective lenses are adjusted to minimise spherical aberrations at visible wavelengths rather than for infrared (Reihani and Oddershede, 2007). Various methods have been used to reduce the effects of aberrations in optical tweezers such as employing deformable mirrors (Ota et al., 2003; Theofanidou et al., 2004), using an SLM (Wulff et al., 2006) and optimisation of the immersion oil (Reihani and Oddershede, 2007). Aberration correction can significantly improve the quality of an optical trap for particle sizes of the order 1 μm or less (Wulff et al., 2006). Optimising the stiffness of an optical trap deep in a sample can extend the range over which measurements can be made and has allowed the trapping of 1 μm diameter polystyrene beads up to a depth of 50 μm (Dienerowitz et al., 2011) and the trapping and spectroscopy of metal nanoparticles (Dienerowitz et al., 2012).

5.4 Sample Holders

Many samples used in optical tweezers are in aqueous solution, where the water provides over-damping to create a stable optical trap. Evaporation of the water can cause problems during measurements as this leads to fluid flow in the sample. To avoid such issues the sample is sealed using a coverglass. Mounting the sample slide with the coverglass towards the objective lens allows focussing deeper into the sample and reduces aberrations. A welled sample slide and a coverglass can be used to contain a sample which can be permanently sealed around the edges of the coverglass using glue. This can allow measurements to be taken on the same sample continuously over a period of hours, where the slide can be used multiple times over several days.

Sometimes fluid flow is desirable and simple sample flow cells can be constructed using a glass slide and two pieces of double-sided tape, mounted to form a channel a few 10 s of μm deep. A coverglass is attached to the top and fluid introduced at one end of the channel using a micropipette.

More sophisticated sample cells can be used based on microfluidic structures and syringe pumps. Optically driven pumps can be implemented in lab on chip devices (Leach et al., 2006; Mushfique et al., 2008a) as well as multipoint viscosity measurements

(Keen et al., 2009) and the 3D mapping of microfluidic flow (Mushfique et al., 2008b).

5.5 Controlling the Trap Position

The simplest method of moving a trapped object in a microscope sample is to move the sample slide relative to the objective lens, and hence, the position of the optical trap. This usually requires keeping the objective lens fixed and controlling the lateral position of the microscope stage. Many different stages are available, having interchangeable sample holders for different applications, and range from simple manually adjustable versions to motorised ones. Many have high-resolution motors and gears and are equipped with encoders, allowing samples to be positioned with a high degree of resolution and repeatability. Sample slide displacements can be performed programmatically and in many cases a constant velocity can be set allowing the spring constant of an optically trapped sphere to be calibrated with the Stokes' drag force. In a similar way to translational displacements, a linear stage and encoder can provide displacements of the objective lens (and hence, focussing of the microscope) relative to the sample slide. Motorised *x*-*y*-*z* positioning systems are available that allow samples to be placed with precisions typically in the 10 s of nm range. Careful consideration is needed when using a motorised stage as backlash in the gears can affect the repeatability of a programmed displacement. In such cases a backlash correction can be applied to the stage controller to minimise these problems. Piezoelectric stages overcome this problem, providing fast and relatively smooth positioning of the sample or objective lens. They are often available with closed loop control and can provide a calibrated displacement readout in microns. However, despite being able to position samples to a precision of better than a nm, their range is limited to approximately 100 μm in each axis. In some applications it is sufficient to control the optical trap position by sample stage movement alone but many applications can also benefit from being able to steer the laser trap over the field of view of the microscope.

5.5.1 *Steering Mirrors (Motorised)*

Moving the optical trap laterally across the field of view of the microscope can be achieved by tilting the direction of the trapping laser beam at the back focal plane of the objective lens. However, due to space restrictions it is not normally convenient, or even possible, to locate beam steering elements at the back aperture in many microscope systems. A common solution is to use a telescope arrangement to achieve beam tilting at a conjugate plane to the back-aperture.

There are various methods of steering traps in optical tweezers systems. The simplest method is to use an adjustable mirror mount as the beam steering component which is positioned in the Fourier plane of the sample. In the case of a modern infinity corrected microscope, a change in angle of the laser beam from the adjustable mirror, and hence a change in angle into the back aperture of the objective, is translated to a change in the lateral position of the optical trap in the sample (see Fig. 5.2a). As with the microscope sample stage, the mirror mount can be manually controlled or driven using motors or piezo actuators. Introducing a second optical trap can be achieved by adding an additional laser to the system. However, a more convenient method is to use a pair of polarising beamsplitters to split and recombine a single laser beam, forming two optical traps that can be positioned within the sample (Atakhorrami et al., 2008; Klein et al., 2007; Moffitt et al., 2006). This approach becomes more difficult for more than two optical traps and hence, for creating large arrays of traps, a time sharing scheme is required where the laser is scanned over multiple positions in the sample plane.

Fast galvanometers allow the trapping laser beam to be scanned rapidly between multiple trap positions, over a large field of view, such that particles do not have time to diffuse away before the next return interval of the laser (Sasaki et al., 1991; Visscher et al., 1993). If the laser is scanned much faster than the characteristic corner frequency of the optical traps then the system behaves as if there are multiple traps present simultaneously, each of a power given by the time-averaged power present at each location. In this case a bandwidth of several kHz is required in order to operate many traps

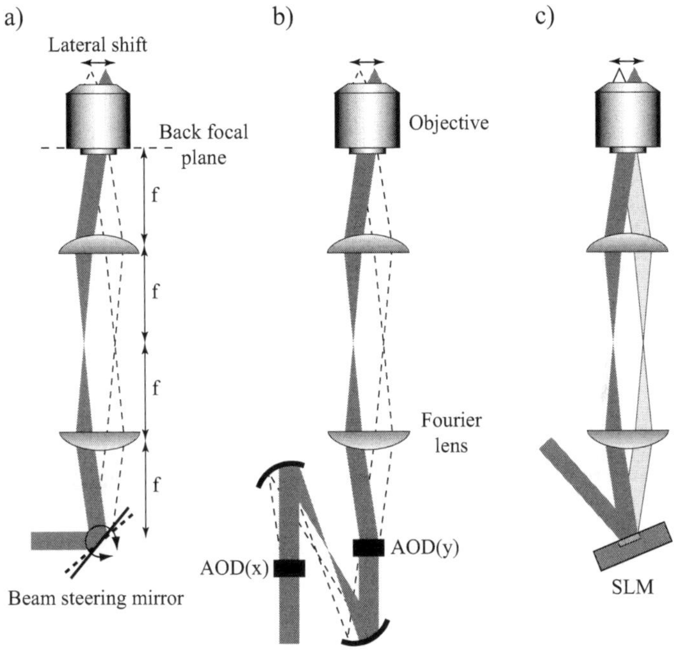

Figure 5.2 Changing the angle of the laser beam into the back of the objective lens results in a lateral translation of the optical trap at the focus. a) Steering the optical trap using a mirror mount or galvo scanning mirror. Multiple traps can be created by time sharing the laser beam between the trapping sites. b) Steering the optical trap in 2D using a pair of orthogonally aligned acousto-optic deflectors. c) Multiple trap generation and steering using a Spatial Light Modulator (SLM). Multiple traps can be created at the focus of the objective lens by calculating an appropriate diffraction pattern to display on the SLM.

in a time-sharing configuration. The mass of the scanning mirrors and their mounts places a limit on the scan speed and hence, for the control of many optical traps, a non-mechanical beam scanning method is preferred (Visscher et al., 1996). Common options include the use of an Acousto-Optic Deflector (AOD) for time sharing traps in a planar geometry (Brouhard et al., 2003; Wallin et al., 2008) or using a Diffractive Optical Element (DOE) for creating multiple traps simultaneously and in arbitrary 3D arrangements (Liesener et al., 2000; Reicherter et al., 1999).

5.5.2 *Acousto-Optic Deflectors*

AODs are fast beam deflectors with which multiple particles can be trapped in an array, achieved by the laser dwelling briefly on each particle before moving to the next. Similar to the case of the scanning mirror approach, the time between scan cycles of the laser must be less than the diffusion time of the particle. AODs are mainly limited to trapping in planar configurations but can operate at up to 10 s of kHz.

A piezo-electric transducer is coupled to one end of an Acousto-Optic (AO) crystal, the other end being angle cut and fixed to an acoustic absorber. The piezo is driven by an RF source operating at 10 s of MHz, inducing an acoustic wave which propagates through the crystal. This acoustic wave creates regions of high and low refractive index along the crystal which acts as a diffraction grating for an incident laser beam. Constant power drives the acoustic wave and adjustment of the RF drive frequency changes the period of the induced grating and hence, the deflection angle of the first-order diffracted beam. A change in the RF drive frequency of Δf results in a change in deflection angle $\Delta\theta_{\text{deflection}}$ given by

$$\Delta\theta_{\text{deflection}} \approx \lambda \frac{\Delta f}{V_{\text{sound}}}, \tag{5.1}$$

where λ is the wavelength of light in air and V_{sound} is the speed of sound in the AO crystal. The time response of the AOD is limited to the time taken for the traveling wave to cross the laser beam diameter, equal to diameter/acoustic speed. For example 5 μs for a 3 mm diameter beam. Figure 5.3 shows a schematic of a typical AOD.

Each AOD creates a single diffraction grating which can be rapidly scanned across the field of view of the microscope. Dual axis arrangements are commonly used where an expanded laser beam is sent through two orthogonally aligned AODs, allowing trapping of multiple objects in a planar configuration (see Fig. 5.2b) (Visscher et al., 1996).

5.5.3 *Spatial Light Modulators*

In recent years there has been a renewed interest in optical tweezers due to the availability of computer controlled Spatial Light

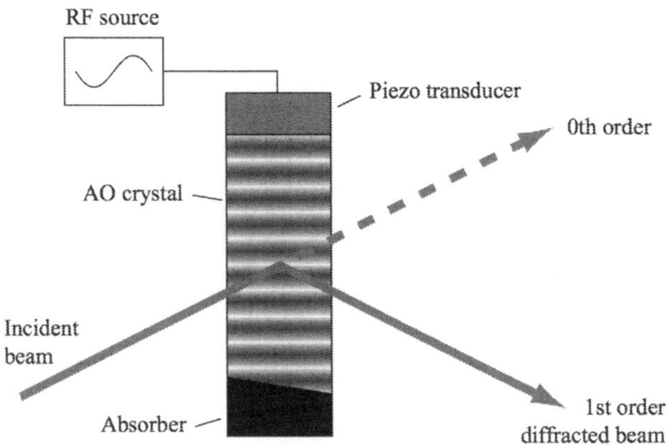

Figure 5.3 Schematic of an Acousto-Optic (AO) deflector. An RF signal drives a piezo-electric transducer which is coupled to one end of an AO crystal, inducing an acoustic wave which modulates the refractive index along the crystal length. Altering the RF drive frequency changes the 1st order diffraction angle of an incident laser beam.

Modulators (SLMs). SLMs are pixellated liquid crystal devices where voltages on each pixel changes the local refractive index of the liquid crystal layer (see Fig. 5.4). When the SLM is illuminated by a laser beam the phase fronts of the reflected beam are shifted by an amount typically between 0 and 2π due to the spatially varying refractive index. They are commonly connected to a computer as an additional monitor display, or sometimes using a dedicated interface card, and are controlled by displaying or loading greyscale images where the pixel values 0 to 255 are mapped to a 0 to 2π phase change. The angle of the incoming beam should be as small as possible (ideally $< 10^\circ$) to ensure that the light interacts with the liquid crystal in the correct orientation. In the case of a lateral shift the SLM modifies the phase of the reflected laser beam in a similar way that a glass prism does, where the phase shift is proportional to the thickness of the prism. Most SLMs, however, are only capable of shifting the phase by an amount up to 2π. Since a phase shift of 2π is the same as a shift of 0, the phase can be wrapped after every interval of 2π. In such systems an SLM is configured as a controllable diffractive optical element allowing a single laser beam to be split, thus forming many independent optical traps (Grier, 2003).

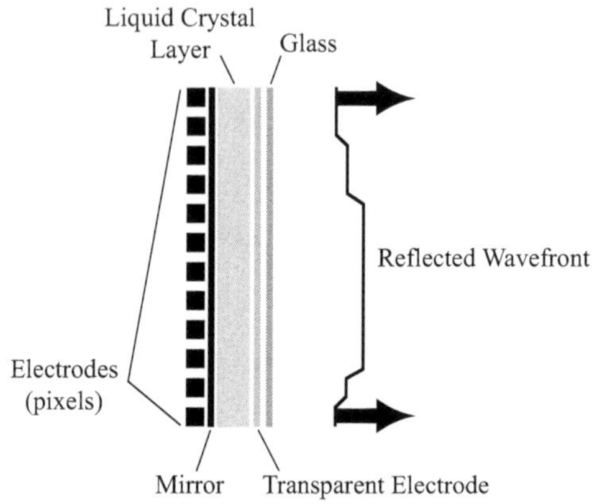

Figure 5.4 A liquid crystal Spatial Light Modulator (SLM). A voltage on a pixel changes the local refractive index of the liquid crystal layer. Phase fronts of a reflected laser beam are shifted typically between 0 and 2π due to the spatially varying refractive index. SLMs are controlled by displaying greyscale images where the pixel values 0 to 255 are mapped to a 0 to 2π phase change of the reflected light.

Standard nematic liquid crystal SLMs are relatively slow compared to other dynamic beam shaping techniques (typically operating with display update rates of 60 Hz). However, SLMs are available that can be updated at frame rates >200 Hz (Preece et al., 2009), allowing rapid switching between diffractive patterns and hence configuration of traps. Unlike the time-sharing configurations using scanning mirrors or AODs, as described above, SLMs allow multiple traps to be created and controlled simultaneously (see Fig. 5.2c). Updating the SLM allows the individual optical traps to be moved independently of each other (Liesener et al., 2000). Such systems are termed "holographic optical tweezers" (Curtis et al., 2002). In addition, an SLM-based system can display holograms that function as additional lenses, creating axial displacements. Remarkably, in such systems the high numerical aperture typically employed in microscope objectives allows the trapping of two objects directly above one another (Bingelyte et al., 2003). Arbitrary displacements can be set independently of each other to create complicated three dimensional structures of micron-sized dielectric spheres (Sinclair et al., 2004a) or living cells (Jordan et al., 2005) using a variety of

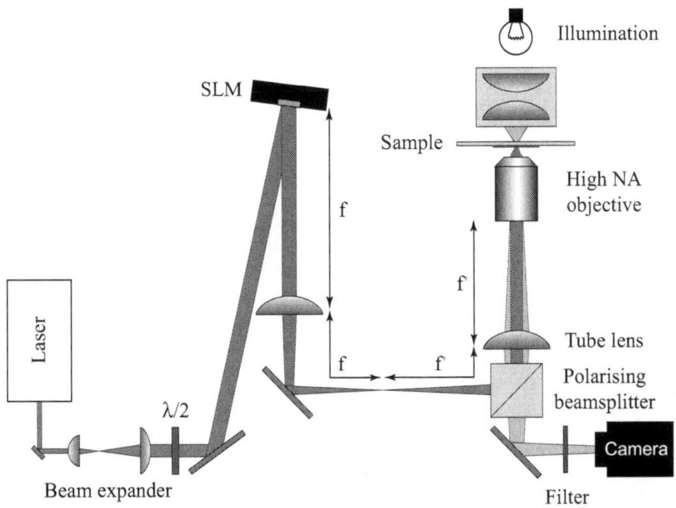

Figure 5.5 A holographic optical tweezers system based around an inverted microscope. A single TEM_{00} laser beam is expanded to fill the aperture of an SLM. A $\lambda/2$ waveplate allows the polarisation of the laser beam to be rotated to the correct orientation for the SLM. SLM control software allows the creation of multiple optical traps, coupled into the microscope using a polarising beamsplitter. The motion of trapped particles in the sample is analysed using a digital video camera.

hologram design algorithms (Leach et al., 2006). Figure 5.5 shows a typical holographic optical tweezers system based around an inverted microscope.

5.5.4 *Hologram Calculation*

5.5.4.1 Gerchberg Saxton

When the SLM is in the Fourier plane of the sample the relationship between the calculated diffraction pattern and the configuration of the traps is a spatial 2D Fourier transform. Various methods exist for defining arbitrary light fields such as those based on the Gerchberg Saxton algorithm (Gerchberg and Saxton, 1972; Sinclair et al., 2004c), which uses Fourier transforms to iteratively transform from the SLM plane to the sample plane and back. In each iteration the amplitude of the fields at the SLM plane and sample planes are replaced by the known amplitude distributions (a Gaussian illumination in the case of the SLM plane and the desired trap arrangement in the case of the sample plane), while the phase is free to evolve in both planes. This algorithm rapidly converges to generate the desired light field (typically 10–20 iterations) despite having a very large number of degrees of freedom. It is useful to note that, when using this method, the phase of each trap is not constrained.

5.5.4.2 Gratings and lenses

In most optical tweezers applications the desired light distribution can be described as a series of diffraction limited spots. In this case one simply sums together the complex holograms for each spot, and displays the phase of the resulting field on the SLM. This simple algorithm is sometimes referred to as "gratings and lenses" and can be easily calculated in real-time for interactive control of a tweezers system (Bowman et al., 2011; Leach et al., 2006). The phase shift, ϕ, can be calculated for each point, (u, v), on the SLM. For a single optical trap centered on the optical axis of the microscope the pattern, $\phi(u, v)_{\text{centered}}$, is just a constant phase. In this case the Fourier lens is all that is required to focus the laser in order to create the optical trap.

$$\phi(u, v)_{\text{centered}} = \text{constant}. \tag{5.2}$$

In order to shift the spot laterally in 1D the pattern is that of a linear diffraction grating which has been blazed to concentrate the diffracted light into the first order. A shift in the x direction is given by

$$\phi(u, v)_{x\text{ shift}} = \left[\left(\frac{ku}{f}\right)x\right] mod 2\pi, \tag{5.3}$$

where $k = 2\pi/\lambda$ is the wavenumber and f is the focal length of the Fourier lens. Extending this to a lateral shift in 2D involves the mod 2π addition of a phase ramp in x with that in y, and is given by

$$\phi(u, v)_{\text{lateral shift}} = \left[\left(\frac{ku}{f}\right)x + \left(\frac{kv}{f}\right)y\right] mod 2\pi. \tag{5.4}$$

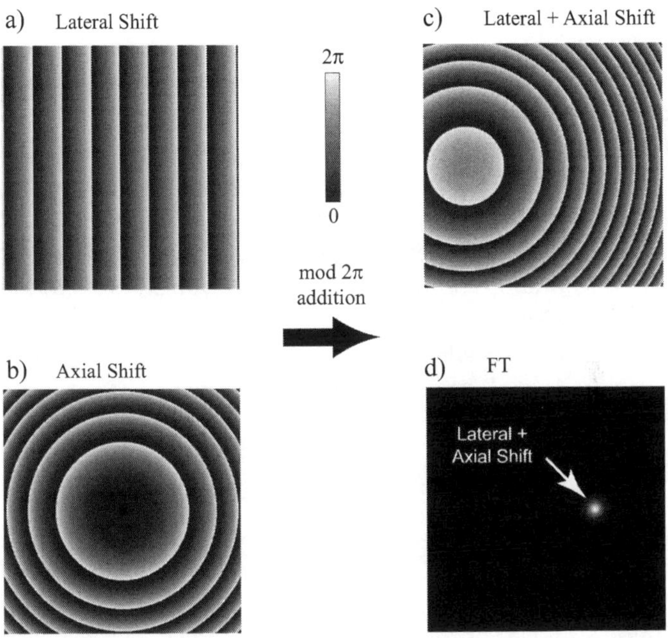

Figure 5.6 Hologram calculation based on a gratings and lenses algorithm. a) Lateral shift due to a linear phase ramp. b) Axial shift due to a Fresnel lens. c) Combination of phase ramp and Fresnel lens to give a lateral and axial shift. d) Fourier transform showing the calculated light field at the focus of the objective lens.

This can be extended to the axial direction also by displaying a Fresnel lens on the SLM. This is given by

$$\phi(u, v)_{\text{axial shift}} = \left[\left(\frac{k(u^2 + v^2)}{2f^2}\right) z\right] mod 2\pi. \tag{5.5}$$

For the general case of a shift in 3D the required hologram is simply the mod 2π addition of the individual holograms for the lateral and axial shifts. Figure 5.6 shows the addition of a lateral and axial shift and can be represented by

$$\phi(u, v)_{3D} = (\phi(u, v)_{\text{lateral}} + \phi(u, v)_{\text{axial}})\, mod 2\pi. \tag{5.6}$$

In addition to changing the lateral position and defocus of the trap, the use of SLMs can also be extended to include the reshaping of the trapping beam. The trapping characteristics of the system can be altered for particular applications or static aberrations can be compensated by the addition of a correction hologram (Wulff et al., 2006). Aberrations can arise due to components within the optical system, sometimes even introduced by the SLM itself.

It is often the case however, that an experiment will require more than one optical trap simultaneously and that the time sharing approaches described previously in the sections above may not be suitable. It is possible to combine multiple holograms into one which will produce multiple traps simultaneously (Grier, 2003; Liesener et al., 2000). Here the individual holograms are expressed as arrays of complex numbers where the argument represents the phase and the magnitude scales with the desired weighting of the spot. To create N traps the complex sum of the individual holograms $\phi(u, v)_n$ is given by

$$\phi(u, v)_{\text{combined}} = arg\left(\sum_{n=1}^{N} A(u, v)_n e^{i\phi(u,v)_n}\right), \tag{5.7}$$

where $A(u, v)_n$ sets the amplitude and phase of each trap. Figure 5.7 shows the complex addition of two holograms to create a laterally displaced spot and a defocussed spot simultaneously.

The SLM is only capable of modulating the phase of the laser beam and not the amplitude. This means that there is often too much light on some parts of the hologram compared to what is required by the hologram calculation, which results in the appearance of unwanted traps or "ghost orders". Ghost orders can be a particular problem when creating large periodic arrays of traps. By using iterative algorithms, for example the Gerchberg Saxton algorithm described above, it is possible to reduce ghost orders to create highly efficient, uniform arrays of traps (albeit at higher computational cost). In this case, the amplitude at the SLM is taken into account when designing the pattern, producing a hologram requiring a more uniform illumination. However, for many typical optical trap configurations, ghost orders can be reduced by applying a slight disordering to the periodic trap positions, thereby reducing the symmetry (Curtis et al., 2005). Even when using highly symmetric arrangements of traps, particulary for small numbers of traps (<10), ghost orders can be reduced by spatially varying the contrast of the

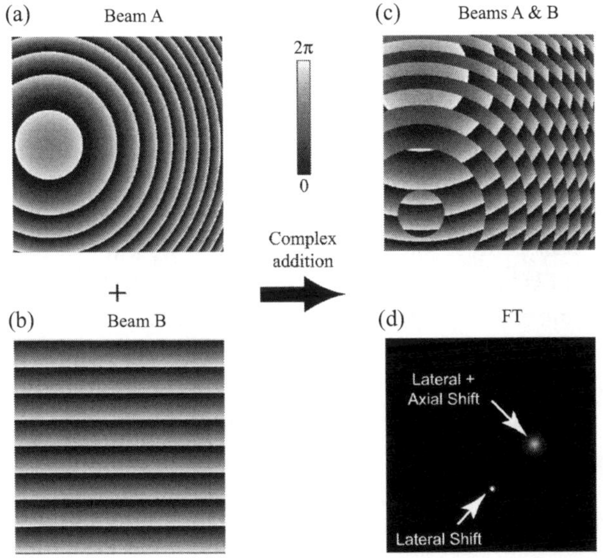

Figure 5.7 Hologram calculation for producing multiple spots. (a) Hologram pattern for producing a lateral and axial shift. (b) Hologram pattern for producing a lateral shift. (c) Complex addition produces a single hologram for creating two spots. (d) Fourier transform showing the calculated light field at the focus of the objective lens.

grating pattern to send unwanted light to the zero order (Bowman et al., 2011; Davis et al., 1999).

In order for holographic optical tweezers to be useful devices in many applications, they need to be able to manipulate objects at a rate which allows interactive control. The "gratings and lenses" algorithm is simple and computationally fast and can be easily performed on a standard PC desktop or laptop computer. However, for applications involving large arrays of trapped objects, or those requiring closed loop control, a faster method is required for generating the holograms. The parallel architecture of many Graphics Processing Units (GPUs) can be exploited to speed up the calculation of holograms by several orders of magnitude (Bianchi and Leonardo, 2010; Preece et al., 2009; Reicherter et al., 2006). Reference (Bowman et al., 2014) describes software for generating these holograms on a consumer GPU, coupled to an easy-to-use interface written in LabVIEW (National Instruments). The GPU engine allows holograms to be calculated at a rate over 300 times faster than using currently available computer processors in a standard desktop PC. This is important when used with the more recent SLMs having higher refresh rates in excess of 200 Hz (Preece et al., 2009).

An important consideration in holographic optical tweezers is the range of control and the positioning resolution of the optical trap. Trap displacements over 10 s of microns is achievable in the lateral plane of the sample (Sinclair et al., 2004b), with a positioning resolution in the nm range (Schmitz et al., 2005). A similar range of control is also achievable in the axial direction when the system has been corrected for aberrations (Dienerowitz et al., 2011).

5.6 Measuring Position and Force

In addition to using optical tweezers to manipulate objects, the position of objects can be accurately monitored and used to determine the forces acting on them (Simmons et al., 1996). Force can be measured in one of two ways. Firstly, the position of the trap is fixed and the displacement of a trapped particle is related to a force by the spring constant, κ, of the trap (valid for small displacements typically $<$ particle radius). Secondly, the trap position is constantly updated to minimise the particle displacement. Both methods require a precise, accurate and high bandwidth measurement of particle position. Optical tweezers have found applications in biology where nanometer resolution position measurements of spherical probes are used to measure femto- to piconewton scale forces (Meiners and Quake, 2000). Tracking microspheres that are optically trapped and suspended in a fluid is an ideal method for studying the localised viscoelastic properties of the fluid (Preece et al., 2011; Tassieri et al., 2012), where the trap is ideal for preventing the particle from diffusing outside the field of view. Optical tweezers can themselves be used to exert forces on the trapped particle which is ideal for active microrheology and in the case of holographic optical tweezers, can be extended to manipulate multiple probes simultaneously.

The position of trapped particles can be measured using either a quadrant photodiode (QPD) (Neuman and Block, 2004) or a high-speed video camera with particle tracking software (Gibson et al., 2008). QPDs are a popular choice for measuring force since they offer precise, high-bandwidth measurements, at up to rates of 100s of kHz. They are, however, normally limited to tracking only one or two particles simultaneously. Video cameras are more convenient, allowing multiple particles to be tracked simultaneously, over large fields of view. It is often useful to record particle positions at a rate faster than the particle's relaxation time. This is the time taken for the autocorrelation of the particle's position to fall to $1/e$, and is typically on the order of milliseconds. Therefore, high-speed video cameras are normally used, allowing measurement rates of several kHz.

For small displacements of the particle (typically less than its diameter) from the center of the trap, the motion of the trapped particle approximates that of a thermally excited, overdamped oscillator in a harmonic potential, characterised by a spring constant or trap stiffness, κ (Berg-Sørensen and Flyvbjerg, 2004; Molloy and Padgett, 2002). The trap can be treated approximately as a linear spring where the restoring force, F, exerted on the particle, in one dimension, is given by

$$F = -\kappa x, \tag{5.8}$$

where x is the particle displacement from trap center. The spring constant is proportional to the trapping laser power and this provides a convenient method of adjusting the characteristics of the trap for different applications. By calibrating the spring constant, it is possible to measure the force exerted on the trapped particle by measuring its displacement from the center of the trap. Various methods exist for calibrating the spring constant and include Stokes' drag, equipartition theorem and power spectrum analysis.

5.6.1 *Quadrant Photodiodes*

Quadrant photodiodes (QPDs) are discrete components consisting of four optically active areas separated by a small gap (typically 10 s of μm). They are used in many applications for detecting the position

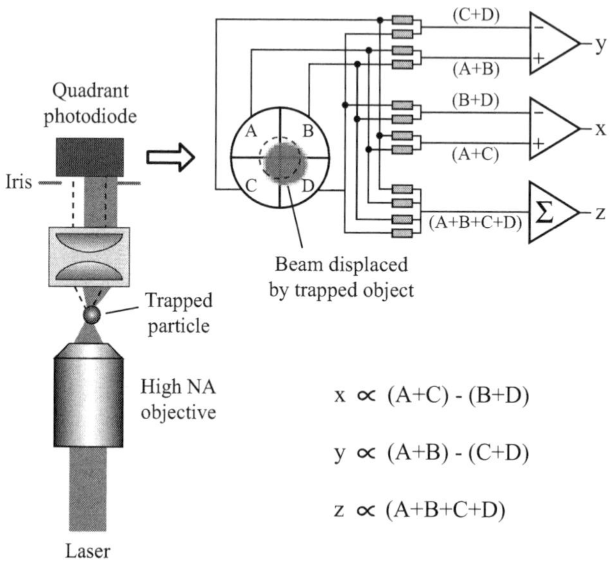

Figure 5.8 A quadrant photodiode located in the back focal plane of the condenser lens can be used to measure the position of a trapped particle. Lateral displacements of the particle results in a lateral shift of the laser spot on the detector, creating a difference in the output signals from the differential amplifiers. Changes in the axial position of the particle results in an increase or decrease in the size of the laser spot, resulting in an increase or decrease in the output signal from the summing amplifier.

of laser beams. In an optical tweezers system, a lens is positioned such that the back focal plane of the condenser is imaged onto the QPD (see Fig. 5.8). Translations of a bead or particle trapped in the laser focus causes translations of the laser spot in the back-focal plane, detected by the QPD. Some systems use a separate laser from that used to trap the object to monitor its position. Figure 5.9 shows a typical configuration of an optical tweezers system with a QPD for particle tracking. Such systems have found applications in colloidal science and biophysics (Finer et al., 1994; Meiners and Quake, 1999).

Axial motions can also be detected using a laser and QPD. Axial displacements change the collimation of the laser and an iris is used to control the maximum capture angle of the condenser (Rohrbach

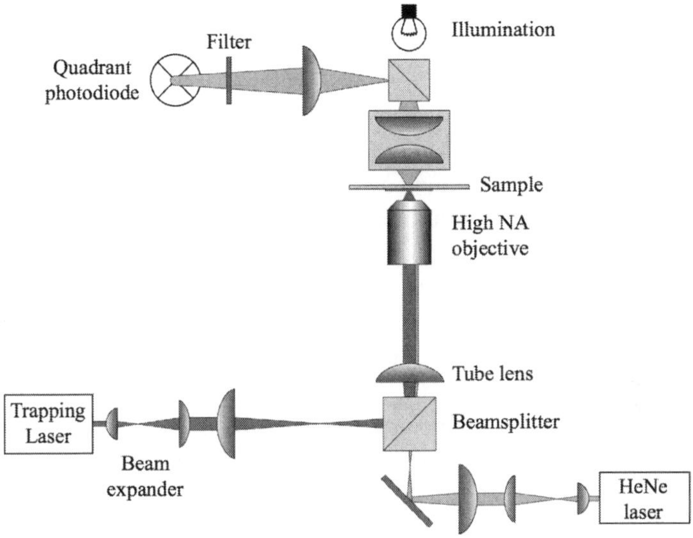

Figure 5.9 Measuring particle position in optical tweezers using a quadrant photodiode. A separate tracking laser can be used to illuminate the sample.

and Stelzer, 2002). Reducing the diameter of the iris improves the axial detection sensitivity but degrades the lateral sensitivity. Careful calibration is required where a similar fixed bead is displaced by known fixed increments in *x*-*y*-*z*, usually using a calibrated motorised or piezo stage, while monitoring the voltage signals from the quadrant photodiode amplifiers.

5.6.2 *Digital Video Cameras*

Digital video cameras allow measurements to be performed over a large field of view compared to QPDs and also allow measurements on multiple objects simultaneously. They are ideally suited for use with holographic optical tweezers where multiple trapped objects can be tracked and used as arrays of force measuring probes. In most cases the camera used to track the particles is the same camera used for viewing the sample. In the case of standard video cameras, the frame rate is normally limited (especially for multiple objects separated in the sample) to ~60 Hz and hence, high-speed video cameras are often used instead. Reference (Keen et al.,

2007) shows similar performance for both a high-speed, full-field, CCD camera and a QPD for displacement measurements in optical tweezers. High-speed cameras have traditionally operated with a buffer allowing only a few seconds of high-speed (several kHz) video to be acquired and then downloaded to the computer. However, high-speed cameras are now available that can acquire video from a reduced field of view at 1 kHz or greater, over long timescales (Di Leonardo et al., 2007a; Gibson et al., 2008; Otto et al., 2010, 2008), where the data can be managed in real-time using a standard desktop PC.

Rather than acquiring the whole image from the camera, a smaller region of interest (ROI) can be chosen such that only the particles under study are imaged. This increases the frame rate of the camera. For each ROI a threshold is applied to reject any background field and the resulting images are used with an algorithm to determine the particle position (see Fig. 5.10). In the case of trapped beads a simple center-of-mass algorithm can be used to determine their lateral position. Usually it is only the bead position data that is stored rather than the raw video data, resulting in faster frame rates and fewer problems with data storage.

Figure 5.10a shows the unprocessed brightfield image of a 2 μm silica sphere. Figure 5.10b shows the image of the bead after a simple threshold has been applied. Pixels that have a value below the threshold are set to zero while those that are above the threshold remain unaltered. The lateral position of the bead is determined by finding the center-of-mass of the thresholded image. This method is simple and fast (typically several kHz). However, a disadvantage of using a simple thresholding technique is that there can be a large jump in the value of pixels on either side of the threshold. In the presence of noise, the outer pixels in the bead image can potentially fluctuate in value by a large amount, resulting in noise in the position measurements. A better method that is less susceptible to noise is to adopt the method shown in Fig. 5.10c where a constant is subtracted from the brightfield image such that the background is zero (any negative pixel values are set to zero). It is possible to determine the position of trapped microspheres to an accuracy of 1/100th of a pixel (Crocker and Grier, 1996) which typically corresponds to a nm range.

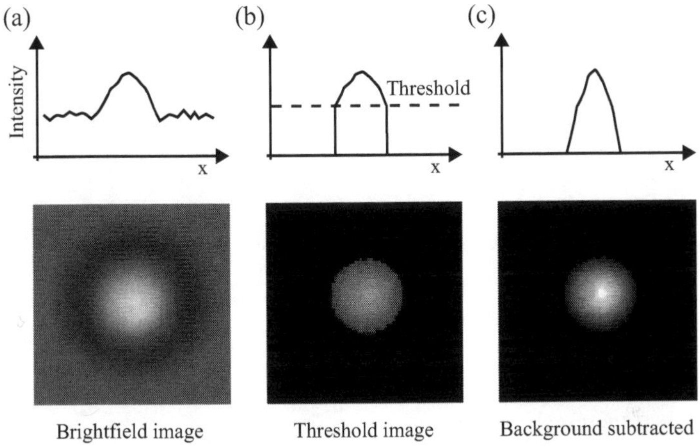

Figure 5.10 Video tracking of a 2 μm silica bead. (a) Unprocessed brightfield image. (b) Thresholded image. (c) Image with background subtracted.

The first step is to calibrate the camera using a sample with known feature sizes. A convenient method is to use a microscope graticule slide having a precision ruling marked on it allowing the camera pixels to be calibrated for position. Graticule slides having rulings with a 10 μm spacing are commonly available. Similarly, a fixed bead on a slide can be displaced by a known amount using a calibrated motorised or piezo sample stage. Once the number of pixels/μm is obtained the center of mass algorithm can provide a real-time position measurement of the bead. Figure 5.11 shows a 1 second sample of the lateral displacement data, x and y, of a 2 μm silica bead. Here the complete data set is obtained from 900,000 measurements acquired at 1.8 kHz, corresponding to 500 seconds.

5.6.3 *Calibration Using Stokes' Drag Method*

A microscopic particle in water is in the low Reynolds number regime, meaning viscous forces dominate over inertial forces. The motion of the particle is overdamped, and the force due to fluid flow can be represented by the Stokes' drag force. The Stokes' drag method enables the spring constant of the optical trap to be determined by dragging a trapped microsphere through a fluid

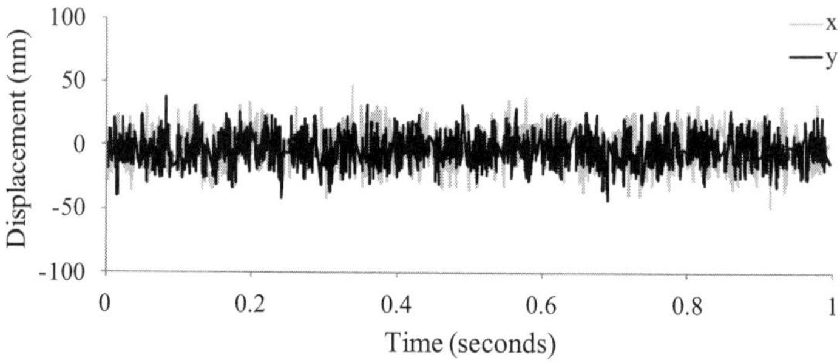

Figure 5.11 Small sample of lateral displacement data for a strongly trapped 2 μm silica bead (κ = 24.3E-6 N/m). Total of 900,000 measurements acquired at 1.8 kHz = 500 seconds. Standard deviation = 13.1 nm. 1 camera pixel represents 152 nm.

at a constant velocity, ν, and measuring its displacement from equilibrium (provided that both the sphere radius and the viscosity of the fluid is known). The hydrodynamic force, F, exerted on the sphere is given by

$$F = -\gamma \nu, \tag{5.9}$$

where γ is the hydrodynamic drag coefficient (assuming the sphere is far from boundaries and other objects) and is related to the radius of the sphere, r, and the viscosity of the fluid, η, by

$$\gamma = 6\pi \eta r. \tag{5.10}$$

Hence, using Eq. 5.8, the spring constant of the trap is given by

$$\kappa = \frac{\gamma \nu}{x}. \tag{5.11}$$

Dragging a trapped probe bead through the fluid can easily be achieved by translating the sample slide at a constant velocity using either a piezo or motorised microscope sample stage. Motorised stages are available that can translate samples at constant velocities in the μm/s range. An important point to note is that this method requires a relatively dilute sample of microspheres in order to avoid the probe bead colliding with other beads, resulting in multiple beads occupying the same optical trap.

5.6.4 *Calibration Using Equipartition Theorem*

Particles immersed in a fluid are constantly bombarded by the atoms or molecules of the fluid, as they undergo random thermal motion. This results in Brownian motion of the particles (Lemons and Gythiel, 1997). When a particle or bead is weakly trapped, its Brownian motion is restricted and it explores a volume around the position of the laser focus. The Equipartition theorem states that for an object in thermal equilibrium, each independent degree of freedom has the same average energy, $k_B T/2$, where k_B is Boltzmann's constant and T is the temperature. This is equal to the average potential energy stored in the trap and hence, for one dimension,

$$\frac{1}{2}k_B T = \frac{1}{2}\kappa\langle x^2\rangle. \tag{5.12}$$

The trap stiffness can be determined without knowledge of the viscosity simply by measuring the variance of the particle's Brownian motion, $\langle x^2\rangle$, for each degree of freedom in a fluid of known temperature, T (Berg-Sørensen and Flyvbjerg, 2004).

5.6.5 *Calibration Using Power Spectrum Analysis*

For a particle weakly held in an optical trap, the particle traces out a constrained Brownian walk. The thermal forces can be balanced against the optical force, the Stokes' drag force and the inertial force to give the following Langevin equation,

$$m\frac{\delta^2 x}{\delta t} + \gamma\frac{\delta x}{\delta t} + \kappa_x x = \zeta(t), \tag{5.13}$$

where m is the mass of the particle and $\zeta(t)$ represents the stochastic force exerted on the particle due to Brownian motion. In the case of overdamped optical tweezers the mass of the trapped particle is insignificant when compared to the viscous forces and hence, this term can be neglected. Taking the square of the Fourier transform of the particle motion, which is governed by the Langevin equation, gives the power spectrum of the thermal motion. The power spectrum has a characteristic Lorentzian shape shown in Fig. 5.12 and is described by the equation

$$PSD = \frac{2k_B T/\pi\gamma}{\omega^2 + \omega_c^2}. \tag{5.14}$$

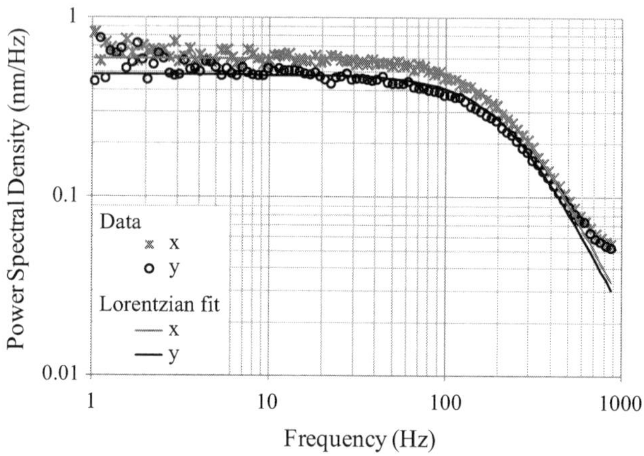

Figure 5.12 Power spectral density. 900,000 position measurements of a 2 μm silica bead was acquired at 1.8 kHz ($\kappa = 24.3E - 6$ N/m). Graph shows a plateau region at low frequency due to the bead's diffusion being limited by the optical trap. The tail, proportional to $1/f^2$, corresponds to free diffusion of the bead over short timescales. The "corner frequency" can be obtained from the graph and is equal to κ/γ.

A common method of calibrating an optical trap fits this equation to the power spectrum of a trapped microsphere and determines its "corner frequency", $\omega_c = \kappa/\gamma$ (Berg-Sørensen and Flyvbjerg, 2004).

5.6.6 *Measuring the Accuracy of Particle Position and Force in Optical Tweezers*

When assessing the performance of a generic measurement system, it is convenient to consider the Allan variance (Allan, 1966) of the sensor output. The Allan variance indicates the timescale over which the system is dominated by Gaussian noise or drift. The precision of position and force measurements using optical tweezers can be obtained by considering the Allan variance of these measurements (Czerwinski et al., 2009; Gibson et al., 2008). This allows characterisation of the stabilities of the measurements and determines the optimal trapping power and measurement, or calibration, time for a particular application.

At short timescales the variation in output is dominated by the noise in the sensor, or the true fluctuations in the system. Both of these fluctuations are Gaussian distributed and hence averaging the data over longer time-windows gives an improvement in the reproducibility of the average value, which improves in proportion to the square root of the number of measurements. At longer timescales, the measurement system is usually subject to drift and averaging the output over these longer time windows ceases to yield any improvement in the reproducibility of the mean value. This transition between the two regimes indicates the useful time over which the system can be said to be stable.

Using the lateral displacement data of a trapped bead, the Allan variance of position can be obtained by

$$\sigma_x^2(\tau) = \frac{1}{2}\langle(x_{n+1} - x_n)^2\rangle, \tag{5.15}$$

where x_n is the average position over the sample period, n, and τ is the time per sample period. Averaging for longer periods should give a better estimate of the mean position of the bead. However, drift results in an increase in the Allan variance for longer averaging times.

Figure 5.13a shows a minimum error of less than 1 nm for an averaging time of 1 second. For averaging times below this value, the drift of the optical system can usually be ignored. The optical force acting on an optically trapped bead can be obtained from a precise measurement of its position relative to the center of the trap. A measure of the trap strength can easily be obtained by monitoring the Brownian motion of the bead prior to the application of the force. Similar to the definition of the Allan variance of position, the equivalent expression for the Allan variance of force is given by

$$\sigma_F^2(\tau) = \sigma_x^2(\tau)\kappa^2, \tag{5.16}$$

Where κ is the spring constant of the trap. The Allan variance of the force measurement of a 2 μm silica bead is shown in Fig. 5.13b. Here it is shown that a force measurement precision of around 20 fN can be achieved for a measurement time of 1 second.

Most of the optical tweezers systems that are designed to measure forces are mounted on an air-isolated optical bench to damp vibrations. The most critical component is the mounting of

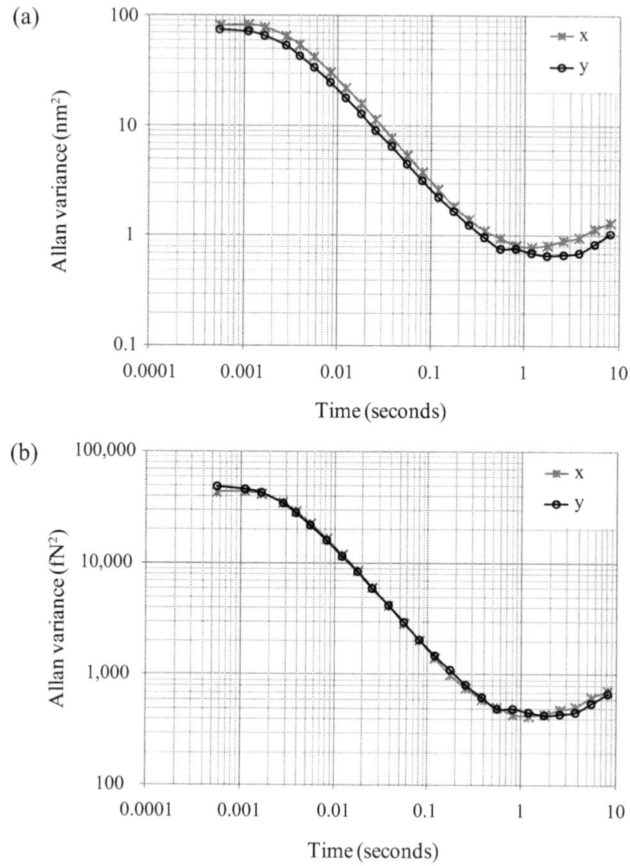

Figure 5.13 Measuring the accuracy of particle position and force. 900,000 position measurements of a 2 μm silica bead trapped in water were acquired at 1.8 kHz ($\kappa = 24.3E - 6$ N/m). (a) Allan variance curves for position measurement. A minimum value less than 1 nm is obtained for a measurement time of 1 second. (b) Similarly the corresponding force curve. A minimum value of around 20 fN is obtained for a measurement time of 1 s.

the objective lens relative to the sample cell. By careful design of the system, the effects of drift can be minimised. Fibre coupling the trapping laser to the microscope, or using materials that have a low coefficient of thermal expansion, can help to achieve a very stable measurement system.

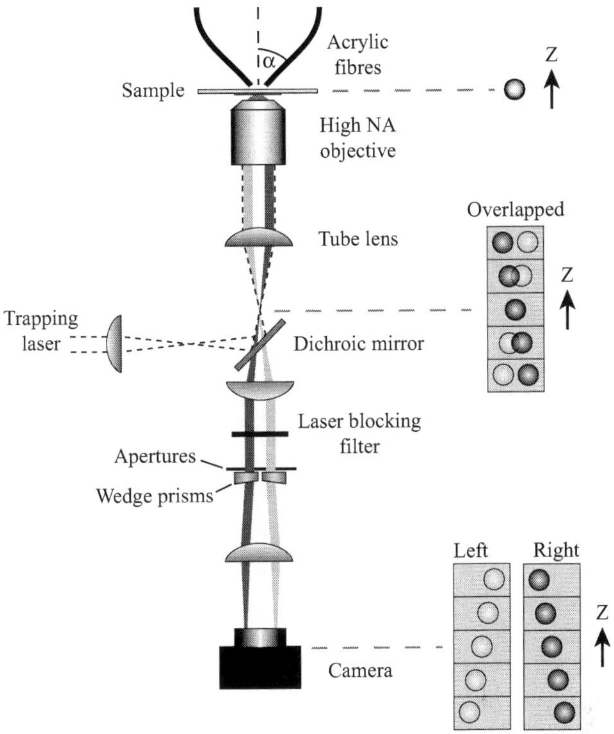

Figure 5.14 An optical configuration used for stereoscopic imaging in optical tweezers. Two angled acrylic fibres (1.5 mm diameter) are used to illuminate the sample. A pair of 1° wedge prisms located in the Fourier plane of the sample separates the stereoscopic images which can be viewed side by side on the same camera.

5.6.7 *Stereoscopic Particle Tracking*

The use of high-speed video in optical tweezers has been shown to be a convenient method for measuring the lateral displacements of particles. However, the image processing required to extract measurements in the axial direction can be computationally demanding. Stereomicroscopy can offer a solution, extending particle tracking from 2D to 3D, which is simple to calibrate and can be performed at high-speed.

Stereomicroscopy was developed to allow perception of depth in microscopes. Each eye is presented with a view of the sample from a different angle and the difference in an object's lateral position in the two views varies with the depth of the object. In optical tweezers the two viewpoints can be obtained using structured illumination of the sample. Dam et. al. uses structured illumination where different colors of light are used for each viewpoint (Dam et al., 2008, 2009). The images are separated using a color camera and then analyzed to obtain the 3D coordinates in real-time.

Figure 5.14 shows a system demonstrated by Bowman et. al. where structured illumination is obtained using two angled optical fibres, each illuminated using high intensity white LEDs (Bowman et al., 2010). The light collected by the objective contains two overlapping images, one from each viewpoint. These images are separated using a pair of 1° wedge prisms located in the Fourier plane of the sample, allowing the images to be viewed side by side on the same camera. In addition to the prisms, a pair of apertures regularises the point spread function of the fibre illuminated image. Lateral motion of an object results in correlated motion between the stereo images, while axial motion displaces each image by the same distance but in opposite directions. Therefore, 3D particle tracking can be performed at high speeds (up to several kHz) as simple 2D tracking on each image is all that is required to extract the axial displacement. 3D tracking resolutions of typically a few nm is possible using this method and it is simple to calibrate. Multiple particles can be tracked simultaneously, which has been used, for example, to measure the 3D orientations of complex microtools for surface scanning (Phillips et al., 2012a).

A variation of this technique replaces the prisms with an SLM, allowing different prisms angles and aperture sizes to be implemented holographically. The SLM can allow the implementation of additional imaging modes such as autofocusing, and darkfield-stereo (Lee et al., 2013). Autofocusing is a useful feature when studying the sedimentation of beads in a fluid (Lee et al., 2014). An optical trap is used to raise the bead tens of μm above the objective focus and then released. The 3D position of the bead is then measured using stereo video particle tracking as it falls under gravity towards the objective. The focal power of a Fresnel lens, implemented on the SLM, is continually updated to keep the bead in focus, thus increasing the axial tracking range beyond that of a standard microscope.

SLM based microscopes can allow ever more complex optical arrangements to be implemented without having the inconvenience of an increasing number of discrete optical components. The stereoscopic microscope can easily be extended from two to four directions by adding additional illumination fibres and updating the hologram displayed on the SLM. This allows two independent sets of 3D measurements to be obtained for tracked objects, where a comparison of these measurements can provide real-time evaluation of the 3D particle tracking accuracy (Hay et al., 2014).

5.7 Conclusions

This chapter has covered some of the many possible configurations of optical tweezers for microrheology. The cost of the optical tweezers system and the stability of position and force measurements are important issues. Optical tweezers can be either constructed using the convenience of commercially available inverted microscopes or custom developed for more specialist applications. The design of optical tweezers ranges from the simple single trap systems to more complex holographic systems for multiple trap control.

The convenience of using high-speed video allows multiple probe beads to be tracked and the forces acting on them measured. Once calibrated, optical tweezers can be used as a tool for microrheology where local viscoelastic properties of fluids can be measured. In addition, they can be used, for example, to accurately position microtools for applying piconewton forces and for tracking complex probes for surface scanning.

Part III

Microrheology

Chapter 6

Introduction to Linear Rheology

Manlio Tassieri
Division of Biomedical Engineering, School of Engineering, University of Glasgow, Glasgow G12 8LT, UK
manlio.tassieri@glasgow.ac.uk

6.1 Introduction

Rheology, from the ancient greek $\rho\epsilon\omega$ (rheō, "flow") + –logy ("study of"), is the study of the flow of matter.

In general, rheology can be described as the study of materials' deformation occurring over a certain period of time. Indeed, even when a material is subject to a finite deformation, it may appear as if it were flowing, as long as the observation time falls within the time window of the process.

Amongst all the possible type of flow that a material could experience, the *shear flow* plays the most important role in the field of rheology. This is defined as the flow induced by the relative motion of two parallel surfaces placed at a constant distance and between which the material is squeezed, as schematically shown in Fig. 6.1. The flow (or deformation as discussed above) is induced by a *shear stress* (σ), which is defined as the ratio between the

Microrheology with Optical Tweezers: Principles and Applications
Edited by Manlio Tassieri

ISBN 978-981-4669-18-4 (Hardcover), 978-981-4669-19-1 (eBook)
www.panstanford.com

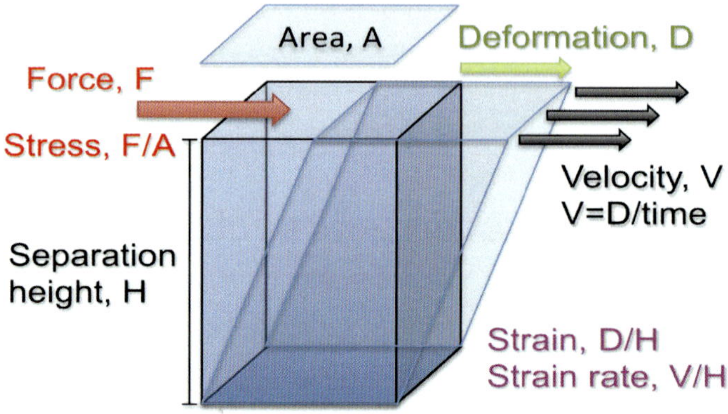

Figure 6.1 A schematic representation of shear flow.

tangential force (F) applied (in this case) to the top surface and the contact area (A) between the latter and the material (having assumed here that the bottom surface is fixed in space). It follows that σ has dimension of a pressure and, with regard to the International System of Units, it is expressed in *pascal* Pa. The deformation of the material is described in terms of *shear strain* (γ), which is given by the ratio between the relative displacement of the two parallel layers in the direction of flow and their relative distance; so that, for the sample in Fig. 6.1: $\gamma = D/H$. The strain is dimensionless. In a similar way, it is possible to describe the shear flow of the material in terms of rate of deformation, i.e. the *shear rate* (or *strain rate*), which is defined as the time derivative of the shear strain ($\dot{\gamma} = \mathrm{d}\gamma/\mathrm{d}t$) and its SI unit is s^{-1}.

The importance of the role played by the shear flow in the field of rheology can be explained by means of two simple thoughts: (i) it is the main flow in fluid processing, this because of the relative motion between the fluid and the inner wall of the pipelines; (ii) most of materials show weaker mechanical properties under shear than others kind of deformations (e.g. compressional or torsional); thus, its importance from a safety point of view. This explains, for instance, why in seismic areas the concrete used for construction is reinforced with iron; so that, buildings do not collapse under shear deformations generated by earthquakes. These considerations, together with its instrumental and analytical simplicity of execution and calculus, make the shear flow an invaluable tool for all rheological studies aimed at uncovering the stress-strain relationships governing the flow of matter.

6.2 Linear Rheology for Simple Shear

According to rheology, all real bodies possess mechanical properties that are intermediate between two *ideal* cases: *perfectly elastic* solid and *purely viscous* fluid.

The first idealisation has been the ground on which *the classical theory of elasticity* has grown and for which, in accordance with Hooke's law, the stress is always directly proportional to strain in 'small' deformation but independent of the rate of strain. With regard to the shear deformation (Fig. 6.1), this translates as:

$$\sigma(t) = G\gamma(t) \tag{6.1}$$

where G is a time-independent elastic constant proportional to the Young's modulus (E) of the material. The dimensional unit of G is the same of a stress (i.e. Pa).

The second idealisation has been the spring from which *the classical theory of hydrodynamics* has flowed and for which, in accordance with Newton's law, the stress is always directly proportional to rate of strain but independent of the strain itself; which in the case of shear flow (Fig. 6.1) writes:

$$\sigma(t) = \eta\frac{\mathrm{d}\gamma(t)}{\mathrm{d}t} \tag{6.2}$$

where η is the time-independent *Newtonian viscosity* of the fluid, whose dimensional unit is $Pa \cdot s$.

In reality, the above two constants of proportionality cannot be considered time-independent anymore and different constitutive equations must be adopted to account the viscoelastic nature of matter. However, in the case of infinitesimal deformations and/or relatively slow flows, the above idealisations represent the asymptotic expressions of the constitutive equations for all real materials. Moreover, the range of stresses (or strains) for which both G and η are invariant, defines the *Linear Viscoelastic* (LVE) regime of a generic material.

Let's revise some fundamental equation describing the stress-strain relationships valid for any generic material in the linear regime. Based on the principle that the effects of sequential changes in strain are additive, the following constitutive equation can be written:

$$\sigma(t) = \int_{-\infty}^{t} G(t-t')\dot{\gamma}(t')\mathrm{d}t' \tag{6.3}$$

where $G(t)$ is the shear *relaxation modulus* of the material and the integration is carried out over all the past times t' up to the current time t. An alternative, but equipollent constitutive equation can be written to express the strain in terms of the history of the time derivative of the stress ($\dot{\sigma}(t) = \mathrm{d}\sigma(t)/\mathrm{d}t$):

$$\gamma(t) = \int_{-\infty}^{t} J(t-t')\dot{\sigma}(t')\mathrm{d}t' \tag{6.4}$$

where $J(t)$ is the shear *creep compliance* of the material, which has dimensions of the inverse of a modulus (Pa^{-1}). For a simple shear measurement, the creep compliance is defined as the ratio of the time-dependent shear strain $\gamma(t)$ to the magnitude σ_0 of the constant shear stress that is switched on at time $t = 0$:

$$J(t) = \frac{\gamma(t)}{\sigma_0} \tag{6.5}$$

For a *perfectly elastic* solid: $J = 1/G$; whereas, for viscoelastic materials $J(t) \neq 1/G(t)$, because of the difference between the two experimental time patterns.

The relaxation modulus and the creep compliance are related by means of a convolution integral:

$$\int_{0}^{t} G(t')J(t-t')\mathrm{d}t' = t \tag{6.6}$$

from which it follows that $G(t)J(t) \leqslant 1$.

It is important to note that Eqs. 6.3, 6.4 and 6.6 are convolution integrals and their mathematical properties turn to be very useful for the solution of rheological problems. In particular, it is worthwhile to remind that the Fourier transform (FT) of a convolution integral equals the point-wise multiplication of the two Fourier-transformed integrand functions. So that, if $f(t)$ and $g(t)$ are two generic functions of the variable t with a convolution integral given by:

$$w(t) = f(t) * g(t) \equiv \int_0^t f(t')g(t-t')\mathrm{d}t' \tag{6.7}$$

its FT writes:

$$\hat{w}(\omega) = \hat{f}(\omega) \cdot \hat{g}(\omega) \tag{6.8}$$

where the symbol "^" indicates the Fourier transform operator, which is defined as:

$$\hat{f}(\omega) = \int_{-\infty}^{+\infty} f(t)e^{-i\omega t}\mathrm{d}t \tag{6.9}$$

where ω is the angular frequency and i is the imaginary unit, i.e. $i^2 = -1$.

When the above maths is applied to Eqs. 6.3, 6.4 and 6.6 they transform as:

$$i\omega\hat{G}(\omega) = \frac{\hat{\sigma}(\omega)}{\hat{\gamma}(\omega)} \tag{6.10}$$

$$i\omega\hat{J}(\omega) = \frac{\hat{\gamma}(\omega)}{\hat{\sigma}(\omega)} \tag{6.11}$$

$$i\omega\hat{G}(\omega) = \frac{1}{i\omega\hat{J}(\omega)} \tag{6.12}$$

The two terms on the left side of Eqs. 6.10 and 6.11 are defined as the *complex modulus* $G^*(\omega) = i\omega\hat{G}(\omega)$ and the *dynamic compliance* $J^*(\omega) = i\omega\hat{J}(\omega)$ of the material; they can also be seen as the Fourier transforms of the time derivative of $G(t)$ and $J(t)$, respectively. They both are complex numbers and can be arbitrarily used to describe the linear viscoelastic properties of materials. In particular, the real and the imaginary parts of the complex modulus $G^*(\omega) = G'(\omega) + iG''(\omega)$ are named as the *storage* and the *loss* moduli and they provide information on the *elastic* and the *viscous* character of the martial, respectively, as described hereafter.

The conventional method of measuring the LVE properties of a material is based on the imposition of an oscillatory stress $\sigma(\omega, t) = \sigma_0 \sin(\omega t)$ (where σ_0 is the amplitude of the stress function) and the measurement of the resulting oscillatory strain, which would have a

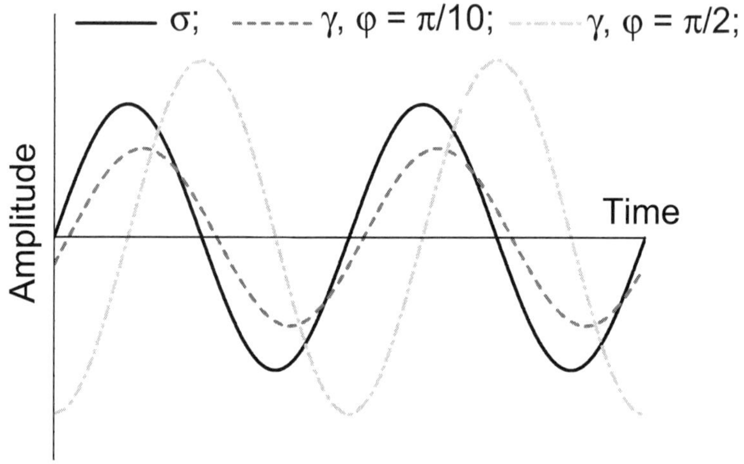

Figure 6.2 Time profile of a simple shear experiment with sinusoidally varying stress and strain.

form like $\gamma(\omega, t) = \gamma_0(\omega)\sin(\omega t - \varphi(\omega))$, where $\gamma_0(\omega)$ is the strain amplitude and $\varphi(\omega)$ is the frequency-dependent phase shift between the stress and the strain, as schematically shown in Fig. 6.2.

Before solving Eq. 6.10 for the above case, it is worthwhile to review some properties of specific trigonometric functions and their Fourier transforms as reported in Table 6.1.

In Table 6.1, ν is the Fourier frequency (instead of ω as in Eq. 6.9, to avoid confusion), δ is the Dirac delta function that on the real number line is zero everywhere except at zero, with an integral of one over the entire real line. Therefore, the delta functions in Table 6.1 are different from zero (i.e. equal to 1) only for $\nu = \pm\omega$. However, since in physics no negative frequencies can exist, the delta function that are *triggered* for negative values of ω are discarded. Moreover, the Fourier transforms of the trigonometric functions

Table 6.1 Fourier transform

Function	Fourier transform
$e^{i\omega t}$	$2\pi\delta(\nu - \omega)$
$\cos(\omega t)$	$\pi[\delta(\nu - \omega) + \delta(\nu + \omega)]$
$\sin(\omega t)$	$-i\pi[\delta(\nu - \omega) - \delta(\nu + \omega)]$

cosine and sine in Table 6.1 have been derived by means of Euler's formula, which states that:

$$e^{ix} = \cos(x) + i\sin(x) \tag{6.13}$$

for any real number x given in radians (in our case $x = \omega t$, with ω expressed in rad/s). Therefore, it follows that:

$$\sin(\omega t) = \frac{\left(e^{i\omega t} - e^{-i\omega t}\right)}{2i} \tag{6.14}$$

$$\cos(\omega t) = \frac{\left(e^{i\omega t} + e^{-i\omega t}\right)}{2}. \tag{6.15}$$

From Eq. 6.10 it follows that the complex modulus is given by the ratio of the Fourier transformed stress and strain functions, which in this case are two sinusoidal functions that transform as it follows:

$$\hat{\sigma}(\omega) = -i\sigma_0\pi\delta(\nu - \omega) \tag{6.16}$$

Whereas, the strain needs some manipulations before being Fourier transformed:

$$\gamma(\omega, t) = \gamma_0(\omega)[\sin(\omega t)\cos(\varphi(\omega)) - \sin(\varphi(\omega))\cos(\omega t)]. \tag{6.17}$$

From which it follows that:

$$\hat{\gamma}(\omega) = -\gamma_0(\omega)\pi\delta(\nu - \omega)[\sin(\varphi(\omega)) + i\cos(\varphi(\omega))]. \tag{6.18}$$

By substituting Eqs. 6.16 and 6.18 into Eq. 6.10 one obtains:

$$G^*(\omega) = \frac{\hat{\sigma}(\omega)}{\hat{\gamma}(\omega)} = \frac{\sigma_0}{\gamma_0(\omega)}\frac{i}{[\sin(\varphi(\omega)) + i\cos(\varphi(\omega))]}. \tag{6.19}$$

From which, by multiplying and dividing the right side of Eq. 6.19 by the conjugate of the denominator, one obtains:

$$G^*(\omega) = \frac{\sigma_0}{\gamma_0(\omega)}\cos(\varphi(\omega)) + i\frac{\sigma_0}{\gamma_0(\omega)}\sin(\varphi(\omega)) \equiv G'(\omega) + iG''(\omega) \tag{6.20}$$

$$|G^*| = \sqrt{G'^2 + G''^2}. \tag{6.21}$$

Equation 6.20 provides the definition of the viscoelastic moduli $G'(\omega)$ and $G''(\omega)$, which provide information on both the elastic and the viscous *nature* of the materials. For example, in the case of a perfectly elastic solid, the stress and the strain are in phase

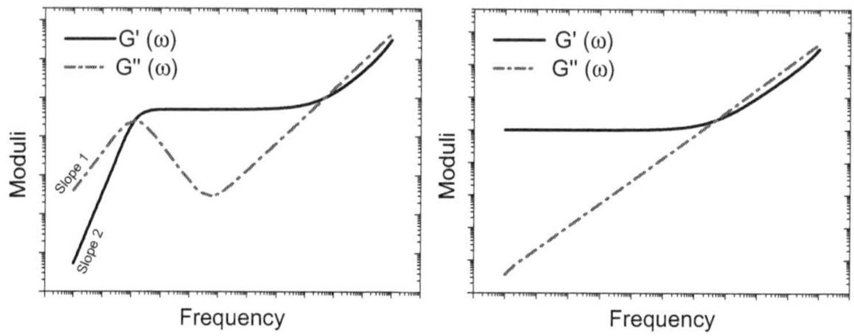

Figure 6.3 Schematic representations of the frequency dependent moduli for a viscoelastic fluid (left) and a complex solid (right). Both the graphs have double logarithmic scales.

and $\varphi(\omega) = 0 \rightarrow G^*(\omega) \equiv G'(\omega)$; whereas, for a purely viscous fluid $\varphi(\omega) = \pi/2 \rightarrow G^*(\omega) \equiv iG''(\omega)$. For complex solids (e.g. gels, rubbers) or viscoelastic fluids (e.g. oil (Hasan et al., 2010), saliva (Stokes and Davies, 2007)) $\varphi(\omega)$ would take any value between the above limits (i.e. $0 \leqslant \varphi(\omega) \leqslant \pi/2$) depending on the frequency at which the stress or the strain is applied, as schematically shown in Fig. 6.3. Moreover, it is important to remind that, in general, for viscoelastic fluids it is always possible to identify a characteristic relaxation time (i.e. the fluid's longest relaxation time), commonly named as τ_D, so that for oscillatory measurements performed at frequencies $\omega \ll \tau_D^{-1}$, both the moduli assume two well defined asymptotic behaviours, respectively: $G'(\omega) \propto \omega^2$ and $G''(\omega) \propto \omega$. The region of frequencies delimited at the top end by τ_D^{-1} takes the name of "*fluid's terminal region*". The abscissa of the low frequency crossover between the two moduli provides a good estimation of τ_D^{-1}.

In a similar way to the calculations made for $G^*(\omega)$, it is possible to derive the expression of $J^*(\omega)$:

$$J^*(\omega) = \frac{\gamma_0}{\sigma_0}\cos(\varphi(\omega)) - i\frac{\gamma_0}{\sigma_0}\sin(\varphi(\omega)) \equiv J'(\omega) - iJ''(\omega) \quad (6.22)$$

where $J'(\omega)$ and $J''(\omega)$ represent the *storage* (elastic) and *loss* (viscous) compliance of the material, respectively.

An alternative but equivalent parameter that is often used instead of both $G^*(\omega)$ and $J^*(\omega)$ for describing the fluids LVE properties is the *complex viscosity* $\eta^*(\omega)$, which is defined as:

$$\eta^*(\omega) = \frac{G^*(\omega)}{i\omega} = \eta'(\omega) - i\eta''(\omega) \tag{6.23}$$

where the individual components of the complex variable are given by:

$$\eta'(\omega) = \frac{G''(\omega)}{\omega} \tag{6.24}$$

$$\eta''(\omega) = \frac{G'(\omega)}{\omega} \tag{6.25}$$

The real component $\eta'(\omega)$ for a viscoelastic liquid approaches the zero-shear viscosity η_0 (i.e., $\eta_0 = \eta(\dot{\gamma})$ for $\dot{\gamma} \to 0$) as the frequency ω approaches zero.

Notice that, $G^*(\omega)$, $J^*(\omega)$ and $\eta^*(\omega)$ are time invariants (Ferry, 1980).

6.3 Simple Mechanical Models of Linear Viscoelastic Behaviour

In the history of rheology, the adoption of mechanical models for describing the materials' LVE properties has been a useful route for the progress of theoretical models capable not only to predict the flow of matter, but also to provide information on materials' molecular structure and interaction.

The analogy between a mechanical model and a real material is build on the assumption that the differential equation correlating strength, elongation and time for the model system is identical to that correlating stress, strain, and time for the specific material under study. In particular, the elastic response of a material (e.g., Eq. 6.1) can be represented by the mechanical elastic deformation of a spring (schematically shown in Fig. 6.4), for which the elongation is always proportional to the force applied. On the other hand, the viscous nature of a material (e.g., Eq. 6.2) can be described by the mechanical behaviour of a viscous damper (a dashpot, schematically shown in Fig. 6.4), for which the rate of elongation is proportional to the force applied. Therefore, in the case of a real material, whose LVE properties fall between the above two ideal behaviours, a generic

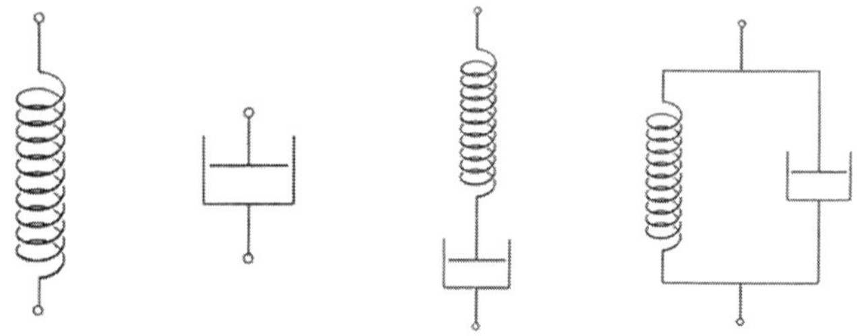

Figure 6.4 Schematic representation of the mechanical elements (elastic spring and dashpot) constituting the Maxwell and the Kelvin-Voigt models.

mechanical model can be obtained by connecting numerous springs and dashpots in a variety of configurations, with no limit on the number of elements that could be use. However, notwithstanding this, the first two possible combinations that one could think of a spring and a dashpot are simply in parallel or in series; indeed, these configurations are known as the Kelvin-Voigt and the Maxwell models, respectively (as shown in Fig. 6.4). They have been the sparks that have led to a myriad of theoretical models seeking to find the stress-strain relationship governing the rheology of complex materials.

The Kelvin-Voigt model is characterised by the fact that both the mechanical elements experience the same amount of deformation γ when a stress σ is applied to the joint extremes. This mathematically translates into the following system of equations:

$$\begin{cases} \gamma = \gamma_E = \gamma_v \\ \sigma = \sigma_E + \sigma_v \end{cases} \tag{6.26}$$

where the subscripts "E" and "v" refer to the elastic (spring) and viscous (dashpot) mechanical elements, respectively. With regards to Eqs. 6.1 and 6.2, the above two equations can be reorganised as:

$$\sigma(t) = G\gamma(t) + \eta\dot{\gamma}(t) \tag{6.27}$$

where the elastic and the viscous stresses have been replaced with the two asymptotic conditions for real viscoelastic materials. Equation 6.27 is a non-homogeneous first order differential equation with constant coefficients (i.e. G and η); whose solution is given by the sum of two terms: (i) the *general* solution of the related homogeneous differential equation and (ii) the *particular* solution of the originally given non-homogeneous differential equation (i.e. Eq. 6.27 with a defined function of $\sigma(t)$). A solution can usually be determined in few mathematical steps, as shown hereafter. The first step is to solve the homogeneous version of it:

$$\eta\dot{\gamma}(t) + G\gamma(t) = 0 \tag{6.28}$$

which can be rearranged as:

$$\eta\dot{\gamma}(t) = -G\gamma(t) \quad or \quad \frac{d\gamma(t)}{dt} = -\frac{G}{\eta}\gamma(t) \tag{6.29}$$

as $\dot{\gamma}(t)$ is the time derivative of $\gamma(t)$. Equation 6.29 can be solved by separation of variables:

$$\frac{d\gamma(t)}{\gamma(t)} = -\frac{G}{\eta}dt \tag{6.30}$$

Integrating both sides of Eq. 6.30 one obtains:

$$Ln(\gamma(t)) = -\frac{G}{\eta}t + C \tag{6.31}$$

where C is a constant that can be determined once the boundary conditions have been specified (e.g. the value of $\gamma(t)$ at time $t = 0$). From Eq. 6.31 it follows that:

$$\gamma(t) = \gamma_0 \exp(-\lambda t) \tag{6.32}$$

where $\gamma_0 = e^C$ and $\lambda = G/\eta$. The latter represents the characteristic relaxation rate of the mechanical system. Note that, Eq. 6.32 satisfy Eq. 6.28 whatever is the value of γ_0.

Let us now solve Eq. 6.27 for the particular and common case of a sinusoidal stress with aptitude σ_0 and frequency ω that is applied at time $t = 0$: $\sigma(t) = \sigma_0 \sin(\omega t)$. In this case, a particular solution ($\gamma_p(t)$) of Eq. 6.27 may look like:

$$\gamma_p(t) = A\sin(\omega t) + B\cos(\omega t) \tag{6.33}$$

where A and B are two constants that can be determined by means of the *method of undetermined coefficients*; as shown hereafter. In order to find these yet unknown coefficients, let's substitute Eq. 6.33 into Eq. 6.27:

$$\begin{aligned}\sigma_0 \sin(\omega t) &= G\gamma_p(t) + \eta\dot{\gamma}_p(t) \\ &= G[A \sin(\omega t) + B \cos(\omega t)] + \eta[A\omega \cos(\omega t) - B\omega \sin(\omega t)]\end{aligned} \tag{6.34}$$

The above equation is true for all t *if and only if* the coefficients of the two functions $\sin(\omega t)$ and $\cos(\omega t)$ are the same on both sides of the equation. Therefore, from Eq. 6.34, one obtains:

$$\begin{cases}\sigma_0 = GA - \eta B\omega \\ GB + \eta A\omega = 0\end{cases} \tag{6.35}$$

from which it follows that:

$$\begin{cases}A = \sigma_0 \frac{G}{G^2 - \eta^2\omega^2} \\ B = \sigma_0 \frac{\eta\omega}{\eta^2\omega^2 - G^2}.\end{cases} \tag{6.36}$$

Therefore, the solution of Eq. 6.27 for the case of an applied sinusoidal stress is:

$$\gamma(t) = \gamma_0 \exp(-\lambda t) + \frac{\sigma_0}{G}\frac{1}{(1-\bar{\omega}^2)}\left[\sin(\omega t) - \bar{\omega}\cos(\omega t)\right] \tag{6.37}$$

where $\bar{\omega} = \omega/\lambda$. Finally, if it is assumed that for all $t \leqslant 0$ the system was undeformed (i.e. $\sigma(t) = \gamma(t) = 0$ for all $t \leqslant 0$), it is then possible to determine γ_0, and Eq. 6.37 becomes:

$$\gamma(t) = \frac{\sigma_0}{G}\frac{1}{(1-\bar{\omega}^2)}\left\{\bar{\omega}\exp(-\lambda t) + [\sin(\omega t) - \bar{\omega}\cos(\omega t)]\right\}. \tag{6.38}$$

At this point it is important to make some considerations on the single exponential decay term in Eq. 6.38, which vanishes for measuring times much longer than the material characteristic relaxation time defined by λ^{-1}. This implies that, for each explored frequency, oscillatory measurements require several cycles before returning a reliable result. Moreover, from a rheological point of view, the number of required cycles can not be established *a priori* (i.e. before performing the measurement), as the viscoelastic behaviour (i.e. G and η) of the sample is *yet unknown* and it actually represents the target of the measurement.

An alternative way to that introduced above for determining the linear viscoelastic properties (i.e., the frequency spectrum) of the Kelvin-Voigt model is to perform the Fourier transform of Eq. 6.27:

$$\hat{\sigma}(\omega) = G\hat{\gamma}(\omega) + \eta i\omega\hat{\gamma}(\omega) = (G + \eta i\omega)\,\hat{\gamma}(\omega). \tag{6.39}$$

Moreover, from Eqs. 6.10 and 6.20 it follows that:

$$G^*(\omega) = \frac{\hat{\sigma}(\omega)}{\hat{\gamma}(\omega)} = G + i\omega\eta. \tag{6.40}$$

Therefore, the Kelvin-Voigt model is a mechanical representation of a viscoelastic material that has a frequency-independent elastic modulus $G'(\omega) = G$ and a viscous modulus that grows linearly with the frequency $G''(\omega) = \omega\eta$; as schematically shown in Fig. 6.5.

The Maxwell model is built on a spring and a dashpot connected in series as shown in Fig. 6.4. In this case, both the mechanical elements experience the same stress, but different deformations. Mathematically, this is described by the following system of equations:

$$\begin{cases} \gamma = \gamma_E + \gamma_v \\ \sigma = \sigma_E = \sigma_v. \end{cases} \tag{6.41}$$

As for the Kelvin-Voigt model, with regards to Eqs. 6.1 and 6.2, the above two equations can be combined as:

$$\gamma(t) = \frac{\sigma(t)}{G} + \frac{1}{\eta}\int \sigma(t)dt + C \tag{6.42}$$

where the elastic and the viscous strains have been replaced with the two asymptotic conditions for real viscoelastic materials. Moreover, the constant C derives from Eq. 6.2, which has been integrated in order to express the strain as function of the stress; however, for simplicity, it is possible to set it egual to zero by assuming the boundary condition that at time $t = 0$ both stress and strain are egual to zero. Equation 6.42 is a non-homogeneous first order integro-differential equation with constant coefficients (i.e. G and η); whose solution is given by the sum of two terms: (i) the *general* solution of the related homogeneous differential equation and (ii) the *particular* solution of the originally given non-homogeneous integro-differential equation (i.e. Eq. 6.42 with a defined function of $\gamma(t)$); as for the Kelvin-Voigt model, a solution can usually be obtained in few mathematical steps, as shown hereafter. The first step is to solve the homogeneous version of Eq. 6.42:

$$\frac{\sigma(t)}{G} + \frac{1}{\eta}\int \sigma(t)dt = 0. \qquad (6.43)$$

This can be differentiated and becomes:

$$\frac{1}{G}\frac{d\sigma(t)}{dt} + \frac{\sigma(t)}{\eta} = 0. \qquad (6.44)$$

Equation 6.44 can be solved by separation of variables:

$$\frac{d\sigma(t)}{\sigma(t)} = -\frac{G}{\eta}dt. \qquad (6.45)$$

Integrating both sides of Eq. 6.45 one obtains:

$$Ln(\sigma(t)) = -\frac{G}{\eta}t + Z \qquad (6.46)$$

where Z is a constant that can be determined once the boundary conditions have been specified. From Eq. 6.46 it follows that:

$$\sigma(t) = \sigma_0 \exp(-\lambda t) \qquad (6.47)$$

where $\sigma_0 = e^Z$ and $\lambda = G/\eta$. The latter represents the characteristic relaxation rate of the Maxwell model. Note that, also in this case, Eq. 6.47 satisfy Eq. 6.43 whatever is the value of σ_0.

Let us now solve Eq. 6.42 for the particular and common case of a sinusoidal strain with aptitude γ_0 and frequency ω that is applied at time $t = 0$: $\gamma(t) = \gamma_0 \sin(\omega t)$. In this case, a particular solution ($\sigma_p(t)$) of Eq. 6.42 may look like:

$$\sigma_p(t) = A\sin(\omega t) + B\cos(\omega t) \qquad (6.48)$$

where A and B are two constants that can be determined by means of the *method of undetermined coefficients*, as shown hereafter. In order to find these yet unknown coefficients, let's substitute Eq. 6.48 into Eq. 6.42:

$$\gamma_0 \sin(\omega t) = \frac{1}{G}[A\sin(\omega t) + B\cos(\omega t)] + \frac{1}{\eta}\int [A\sin(\omega t) + B\cos(\omega t)]dt \qquad (6.49)$$

In order to eliminate the integral term, let us to differentiate it:

$$\omega\gamma_0\cos(\omega t) = \frac{\omega}{G}[A\cos(\omega t) - B\sin(\omega t)] + \frac{1}{\eta}[A\sin(\omega t) + B\cos(\omega t)] \quad (6.50)$$

The above equation is true for all t *if and only if* the coefficients of $\sin(\omega t)$ and $\cos(\omega t)$ are the same on both sides of the equation. Therefore, by comparing the coefficients of the above two functions in Eq. 6.50, one obtains:

$$\begin{cases} \omega\gamma_0 = \frac{\omega A}{G} + \frac{B}{\eta} \\ \frac{A}{\eta} - \frac{\omega B}{G} = 0 \end{cases} \quad (6.51)$$

from which it follows that:

$$A = \frac{\bar{\omega}^2 G\gamma_0}{(1+\bar{\omega}^2)} \quad \text{and} \quad B = \frac{\bar{\omega} G\gamma_0}{(1+\bar{\omega}^2)} \quad (6.52)$$

Therefore, the solution of Eq. 6.42 for the case of an applied sinusoidal strain is:

$$\sigma(t) = \sigma_0\exp(-\lambda t) + \frac{\bar{\omega} G\gamma_0}{(1+\bar{\omega}^2)}[\bar{\omega}\sin(\omega t) + \cos(\omega t)] \quad (6.53)$$

Finally, if it is assumed that for all $t \leqslant 0$ the system was undeformed (i.e. $\sigma(t) = \gamma(t) = 0$ for all $t \leqslant 0$), it is then possible to determine σ_0, and Eq. 6.53 becomes:

$$\sigma(t) = \frac{\bar{\omega} G\gamma_0}{(1+\bar{\omega}^2)}[\bar{\omega}\sin(\omega t) + \cos(\omega t) - \exp(-\lambda t)] \quad (6.54)$$

Coherently with the solution found for the Kelvin-Voigt model, also in this case, it is interesting to notice the presence of a single exponential decay term in Eq. 6.54; which vanishes for measuring times much longer than the material characteristic relaxation time defined by λ^{-1}. Therefore, for each explored frequency, in order to perform a reliable oscillatory measurement, it is required a long measuring time, which cannot be known before performing the measurement.

In the frequency domain, the maths simplifies and the number of steps required to define the linear frequency spectrum of the Maxwell model reduces significantly. Indeed, by performing the Fourier transfer of the total strain one obtains:

$$\hat{\gamma}(\omega) = \hat{\gamma}_E(\omega) + \hat{\gamma}_v(\omega) \quad (6.55)$$

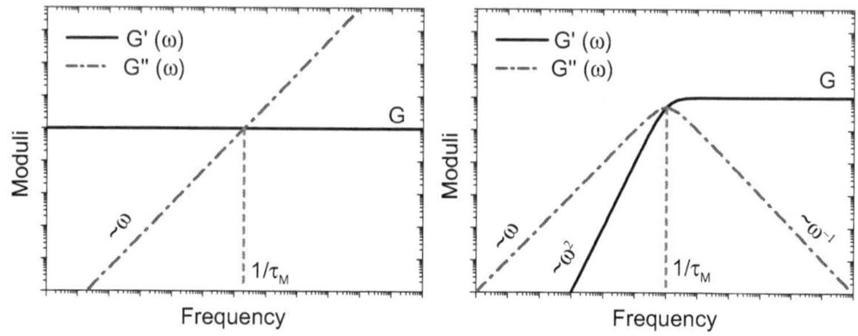

Figure 6.5 Schematic representations of the frequency dependent moduli for the Kelvin-Voigt mechanical model (left) and for the Maxwell model (right). Both the graphs have double logarithmic scales. Where $\tau_M \equiv \lambda^{-1} = \eta/G$ is the characteristic relaxation time of the related mechanical model.

Yet, from Eqs. 6.1 and 6.2 it follows that:

$$\hat{\gamma}(\omega) = \frac{\hat{\sigma}(\omega)}{G} + \frac{\hat{\sigma}(\omega)}{i\omega\eta} = \left(\frac{1}{G} + \frac{1}{i\omega\eta}\right)\hat{\sigma}(\omega) \tag{6.56}$$

Moreover, from Eqs. 6.10 and 6.20 one can write:

$$G^*(\omega) = \frac{\hat{\sigma}(\omega)}{\hat{\gamma}(\omega)} = \frac{i\omega\eta}{1 + i\omega\tau_M} = \frac{\tau_M\eta\omega^2}{1+\omega^2\tau_M^2} + i\frac{\eta\omega}{1+\omega^2\tau_M^2} \tag{6.57}$$

where $\tau_M \equiv \lambda^{-1} = \eta/G$ is the characteristic relaxation time of the mechanical model. Therefore, the Maxwell model is a mechanical representation of a viscoelastic material having both the moduli dependent by the frequency as schematically shown in Fig. 6.5.

Having introduced both the models, it is interesting to highlight the rheological meaning of τ_M in both the cases. Indeed, λ identifies a characteristic frequency that delimits two frequency regions where the elastic and the viscous nature of the material dominate on each other, respectively. In the Maxwell model, a total of four asymptotic behaviours of the material's viscoelastic properties can be identified: i.e., (i) for $\omega < \lambda$ and in the limit of $\omega \rightarrow 0$ both the moduli tend to vanish asymptotically to $(\eta\omega)^2/G$ for $G'(\omega)$ and to $\eta\omega$ for $G''(\omega)$, with the viscous character of the material dominating on the elastic one: $G''(\omega) > G'(\omega)$. Whereas, (ii) for $\omega > \lambda$ and in the limit of $\omega \rightarrow +\infty$ the elastic modulus tends to G and the viscous modulus tends to vanish asymptotically to $G^2/(\omega\eta)$, leaving the elastic component ruling the rheology of the material. On the other hand, for the Kelvin-Voigt model: (i) for $\omega < \lambda$ the elastic nature of the material dominates on the viscous one ($G'(\omega) \equiv G > G''(\omega) \equiv \eta\omega$); whereas, (ii) for $\omega > \lambda$ the viscous nature of the material dominates on the elastic one ($G''(\omega) \equiv \eta\omega > G'(\omega) \equiv G$).

Of course not all the viscoelastic materials can be represented by the above two mechanical models, actually very few. However, all real materials can be modelled with either one of the *generalised* versions of them. Specifically, the viscoelastic properties of a complex material can be modelled by a mechanical system made of a *sufficiently* large number of Kelvin-Voigt or Maxwell systems connected in series or in parallel; where each system contributes to describe a particular relaxation time τ_i of the material by means of the two fundamental parameters G_i and η_i: $\tau_i = \eta_i/G_i$.

Chapter 7

Statistical Mechanics and Diffusion Processes

Adrian Baule
School of Mathematical Sciences, Queen Mary University of London, London E1 4NS, UK
a.baule@qmul.ac.uk

7.1 Introduction

The goal of a microrheological experiment is to infer the viscoelastic properties of the material (e.g., a complex fluid) from a measurement of the trajectories of an immersed micron-sized bead. The size of the bead is large compared with the size of individual molecules of the fluid (a few nanometres), but still small enough so that the bead experiences the effect of thermal fluctuations of the fluid. In the absence of any external driving the bead is then in thermal equilibrium with the surrounding fluid, which acts as a heat reservoir. As a consequence, the motion of the bead is governed by the laws of statistical mechanics prescribing, e.g., precise probabilities for the bead's position and velocity. In this chapter, we discuss the statistical mechanical foundations of microrheology. We develop in particular the mathematical formalism that allows to

Microrheology with Optical Tweezers: Principles and Applications
Edited by Manlio Tassieri

ISBN 978-981-4669-18-4 (Hardcover), 978-981-4669-19-1 (eBook)
www.panstanford.com

describe the dynamics of the bead and derive relations between the quantities of interest and the directly measurable observables.

This chapter is organized as follows. In Sections 7.2–7.4 we focus on the case of simple viscous fluids, for which the formalism can be developed starting from first principles. In Section 7.2 we introduce the concept of Langevin equations that are used to describe diffusion processes in fluids. Sections 7.3 and 7.4 are more mathematical in nature, discussing the calculation of probability density functions from Langevin equations and the overdamped limit, in which the bead's inertia can be neglected. In Section 7.5 we finally introduce generalised Langevin equations that can capture the dynamics of a bead in a viscoelastic medium.

7.2 Diffusion Processes

We are interested in describing the motion of a particle of mass m in a viscous fluid at thermal equilibrium. If the particle's position is denoted by $X(t)$, Newton's 2nd law states that $m\ddot{X}(t)$ equals the sum of all the forces acting on the particle. In the following, we assume three types of forces that are sufficiently general to model a variety of different scenarios at equilibrium:

(1) A friction force due to the viscosity of the fluid environment.
(2) A conservative force originating from a potential U induced, e.g., by optical tweezers.
(3) A thermal force due to the heat fluctuations of the fluid environment, which is held at some temperature T.

Overall, we obtain the equation of motion

$$m\ddot{X}(t) = -\gamma\dot{X}(t) - U'(X(t)) + \xi(t), \tag{7.1}$$

which describes the motion of the particle for a given initial velocity v_0 and initial position x_0. For simplicity, we restrict our discussion to one dimension. In Eq. 7.1 the friction force is assumed linear in $\dot{X}(t)$. The friction coefficient γ is related to the radius of the particle a and the viscosity of the fluid η via Stokes' law

$$\gamma = 6\pi\eta a, \tag{7.2}$$

which is characteristic for a Newtonian fluid. Ultimately, Eq. 7.2 will allow us to infer η from the properties of the observed particle position $X(t)$. The conservative force in Eq. 7.1 is just $-U'(X(t))$, while the thermal force $\xi(t)$ represents the effect of the thermal environment. Physically, $\xi(t)$ models the forces due to the collisions of the particle with the fluid molecules, which have an average kinetic energy $\propto k_B T$. Since we assume that the particle is several orders of magnitude larger than the fluid molecules and the collisions act in a highly irregular way, we do not resolve the dynamics of the fluid molecules, but instead describe $\xi(t)$ as a *random force*. This random force models essentially the thermal fluctuations of the fluid environment. Even in the absence of a potential, the particle will thus move in the fluid under this force, undergoing a diffusive motion. Such a motion is also called *Brownian motion* and the particle undergoing it is called a Brownian particle. In the presence of a potential, $X(t)$ described by Eq. 7.1 is in general called a diffusion process.

Clearly, $X(t)$ will also be a random variable due to the dependence on $\xi(t)$: The value $X(t)$ is different every time we repeat the dynamical evolution of the particle, even for the same initial conditions of Eq. 7.1. Equation 7.1 is thus strictly speaking a stochastic differential equation, which is also called a *Langevin equation* in a physics context. Much of the discussion in this chapter will deal with Langevin equations and how to obtain information on the properties of $X(t)$. The random character of $X(t)$ highlights that the only useful information is contained in average quantities such as, e.g., the average position, which is obtained from the measurement of a large number of trajectories with identical initial conditions.

We generally denote averages with respect to the realizations of the diffusion process by brackets $\langle \ldots \rangle$. This average is also called an *ensemble average*. For clarity, let us consider a function g of the particle position $X(t)$. The average $\langle g(X(t)) \rangle$ is obtained in an experiment as

$$\langle g(X(t)) \rangle = \frac{1}{N} \sum_{i=1}^{N} g(x_i), \tag{7.3}$$

where x_i is the result for $X(t)$ in the ith measurement, repeated N times starting with v_0, x_0. In order for Eq. 7.3 to be a good estimate of the true average, N needs to be sufficiently large (see the discussion below in Section 7.2.4.1) and only converges for $N \to \infty$ to the true average. In this way, we obtain, e.g., the average position $\langle X(t) \rangle$ by setting $g(x) = x$. By considering higher moments $g(x) = x^m$ one can obtain more detailed information on the stochastic process. Formally, the properties of X at any given time t are contained in the probability density function (pdf) $f_X(x, t)$: the probability to observe a value of $X(t)$ in the interval $[x, x + \mathrm{d}x]$ at the time t is given by $f_X(x, t)\mathrm{d}x$.

$$\langle g(X(t)) \rangle = \int_{-\infty}^{\infty} \mathrm{d}x \, g(x) f_X(x, t). \tag{7.4}$$

Note that $f_X(x, t)$ also depends on the initial conditions v_0, x_0.

In the next subsections, we investigate $\xi(t)$ in more detail and show in particular, how moments of $X(t)$ can be calculated from the statistics of $\xi(t)$ without resorting to the pdf $f_X(x, t)$. The calculation of probability density functions is discussed in Section 7.3.

There are a number of excellent textbooks that provide an introduction to the theory of diffusion processes, most notably Refs. (Gardiner, 2010; Risken, 1996; Van Kampen, 2007) on which the discussion in Sections 7.2–7.4 is mostly based.

7.2.1 *Velocity of a Brownian Particle*

The simplest physically relevant special case of Eq. 7.1 is free diffusion without a potential force: $U(x) = 0$. In this case, we can reduce the problem to a single degree of freedom, which is the particle's velocity. We set $V(t) = \dot{X}(t)$ and obtain from Eq. 7.1 the equation of motion for $V(t)$

$$m\dot{V} = -\gamma V + \xi(t), \tag{7.5}$$

for an initial velocity $V(0) = v_0$. This model describes the velocity of a Brownian particle, first introduced by Ornstein and Uhlenbeck (Uhlenbeck and Ornstein, 1930). Here, we can interpret the ratio $\tau_r = m/\gamma$ as a characteristic time scale of the particle's relaxation: In the absence of any forcing the particle's velocity relaxes as $v_0 e^{-t/\tau_r}$.

Since Eq. 7.5 is a linear first order differential equation with a time-dependent force term $\xi(t)$, we can solve Eq. 7.5 directly for $V(t)$

$$V(t) = v_0 e^{-t/\tau_r} + e^{-t/\tau_r} \int_0^t \mathrm{d}t'\, e^{t'/\tau_r} \frac{1}{m} \xi(t') \mathrm{d}t'. \quad (7.6)$$

Before we can use Eq. 7.6 to calculate averages of $V(t)$, we need to infer the properties of the thermal force $\xi(t)$ from physical considerations:

(1) The thermal force does not provide a systematic bias, since we know from the observation of the free diffusion of a Brownian particle that there is no net velocity in an equilibrium state. In general, the dynamics of the particle has settled into an equilibrium state, when the averages of any observables have become stationary. If the particle is initially in an out-of-equilibrium situation, e.g., it starts with some constant velocity v_0, stationarity is achieved for $t \to \infty$. This implies that the average velocity at equilibrium $\langle V \rangle_{\mathrm{eq}}$ is

$$\lim_{t\to\infty} \langle V(t) \rangle = \langle V \rangle_{\mathrm{eq}} = 0. \quad (7.7)$$

Equation 7.7 is essentially a statement of the 2nd law of thermodynamics: Thermal fluctuations alone are not able to generate a directed motion on average, which, in turn, could be used to generate mechanical work. The consequence for $\xi(t)$ is that we need

$$\langle \xi(t) \rangle = 0. \quad (7.8)$$

We obtain then from Eq. 7.6

$$\begin{aligned} \langle V(t) \rangle &= v_0 e^{-t/\tau_r} + e^{-t/\tau_r} \int_0^t \mathrm{d}t'\, e^{t'/\tau_r} \frac{1}{m} \langle \xi(t') \rangle \\ &= v_0 e^{-t/\tau_r}, \end{aligned} \quad (7.9)$$

which satisfies Eq. 7.7. Note that Eq. 7.9 follows from the properties of the average Eq. 7.3, which is a linear operation.

(2) An empirical observation shows that the trajectory of a Brownian particle is highly irregular with sudden changes of velocity. This implies that the random force has an extremely short correlation time, which can be expressed mathematically in terms of the covariance

$$\langle \xi(t)\xi(t') \rangle = \Gamma\delta(t - t'), \quad (7.10)$$

where $\delta(t)$ denotes the Dirac delta function and the parameter Γ is called the noise intensity or noise strength. The noise intensity is determined by the properties of thermal equilibrium (see below). Equations 7.8 and 7.10 characterize ξ in terms of its mean and covariance.

Side remark: For a full specification of $\xi(t)$ we also require information on higher order statistics at different times. This is necessary, in particular, in order to derive the full pdf of the stochastic process. We assume that $\xi(t)$ is a Gaussian random variable, so that the n-point correlation function is zero for odd n and for even n given by

$$\begin{aligned}\left\langle \xi(t_1)\xi(t_2)\cdots\xi(t_n)\right\rangle &= \sum_{\text{perm}} \left\langle \xi(t_{j_1})\xi(t_{j_2})\right\rangle \cdots \left\langle \xi(t_{j_{n-1}})\xi(t_{j_n})\right\rangle \\ &= \Gamma^{n/2} \sum_{\text{perm}} \delta(t_{j_1} - t_{j_2}) \cdots \delta(t_{j_{n-1}} - t_{j_n}), \end{aligned} \tag{7.11}$$

where the sum is taken over all permutations of $j_1, \ldots, j_n$ that lead to distinct products $\left\langle \xi(t_{j_1})\xi(t_{j_2})\right\rangle \cdots \left\langle \xi(t_{j_{n-1}})\xi(t_{j_n})\right\rangle$. For $n = 4$ we obtain for example

$$\begin{aligned}\left\langle \xi(t_1)\xi(t_2)\xi(t_3)\xi(t_4)\right\rangle &= \Gamma^2(\delta(t_1 - t_2)\delta(t_3 - t_4) \\ &+ \delta(t_1 - t_3)\delta(t_2 - t_4) + \delta(t_1 - t_4)\delta(t_2 - t_3)). \end{aligned} \tag{7.12}$$

(3) Since the stochastic velocity process $V(t)$ takes place at thermal equilibrium, the particle's average kinetic energy needs to satisfy the *equipartition theorem* in the stationary regime

$$\langle E_{\text{kin}}\rangle_{\text{eq}} = \frac{1}{2}m\left\langle V^2\right\rangle_{\text{eq}} = \frac{1}{2\beta}, \tag{7.13}$$

where β is the inverse temperature: $\beta^{-1} = k_{\text{B}}T$. In order to understand the implication of Eq. 7.13 on the noise strength Γ, we need to calculate $\left\langle V^2\right\rangle_{\text{eq}}$. Just like the average velocity at equilibrium, we obtain $\left\langle V^2\right\rangle_{\text{eq}}$ in the limit $t \to \infty$: $\left\langle V^2\right\rangle_{\text{eq}} = \lim_{t\to\infty}\left\langle V^2(t)\right\rangle$. In the following, we calculate $\left\langle V^2(t)\right\rangle$ by first considering the covariance $\left\langle V(t_1)V(t_2)\right\rangle$ using the solution Eq. 7.6. Multiplying two solutions $V(t_1)$ and $V(t_2)$ with Eq. 7.6

and taking the average leads to

$$\begin{aligned}\left\langle V(t_1)V(t_2)\right\rangle &= v_0^2 e^{-(t_1+t_2)/\tau_r} + e^{-(t_1+t_2)/\tau_r} \\ &\times \int_0^{t_1} \mathrm{d}t_1' \int_0^{t_2} \mathrm{d}t_2'\, e^{(t_1'+t_2')/\tau_r} \frac{1}{m^2} \left\langle \xi(t_1')\xi(t_2')\right\rangle, \end{aligned} \tag{7.14}$$

where we have neglected terms linear in $\xi(t)$ due to Eq. 7.8. With Eq. 7.10 we can calculate the double integral as follows

$$\begin{aligned}\left\langle V(t_1)V(t_2)\right\rangle &= v_0^2 e^{-(t_1+t_2)/\tau_r} + e^{-(t_1+t_2)/\tau_r} \\ &\quad \times \int_0^{t_1} \mathrm{d}t_1' \int_0^{t_2} \mathrm{d}t_2'\, e^{(t_1'+t_2')/\tau_r} \frac{\Gamma}{m^2} \delta(t_1'-t_2') \\ &= v_0^2 e^{-(t_1+t_2)/\tau_r} + \frac{\Gamma}{m^2} e^{-(t_1+t_2)/\tau_r} \\ &\quad \times \Theta(t_2-t_1) \int_0^{t_1} \mathrm{d}t_1'\, e^{2t_1'/\tau_r} \\ &\quad + \frac{\Gamma}{m^2} e^{-(t_1+t_2)/\tau_r} \Theta(t_1-t_2) \int_0^{t_2} \mathrm{d}t_2'\, e^{2t_2'/\tau_r} \\ &= v_0^2 e^{-(t_1+t_2)/\tau_r} + \frac{\Gamma\tau_r}{2m^2} \Theta(t_2-t_1) \\ &\quad \times \left(e^{-(t_2-t_1)/\tau_r} - e^{-(t_1+t_2)/\tau_r}\right) + \\ &\quad + \frac{\Gamma\tau_r}{2m^2} \Theta(t_1-t_2) \left(e^{-(t_1-t_2)/\tau_r} - e^{-(t_1+t_2)/\tau_r}\right) \\ &= \left(v_0^2 - \frac{\Gamma\tau_r}{2m^2}\right) e^{-(t_1+t_2)/\tau_r} + \frac{\Gamma\tau_r}{2m^2} e^{-|t_2-t_1|/\tau_r}. \end{aligned} \tag{7.15}$$

Here, $\Theta(t)$ denotes the Heaviside step function. Therefore, setting $t = t_1 = t_2$ we have

$$\left\langle V(t)^2\right\rangle = v_0^2 e^{-2t/\tau_r} + \frac{\Gamma\tau_r}{2m^2}\left(1 - e^{-2t/\tau_r}\right). \tag{7.16}$$

The statement of equipartition is thus

$$\lim_{t\to\infty} \left\langle V^2(t)\right\rangle = \left\langle V^2(t)\right\rangle_{\mathrm{eq}} = \frac{\Gamma\tau_r}{2m^2} = \frac{1}{m\beta}, \tag{7.17}$$

or

$$\Gamma = \frac{2m}{\tau_r\beta} = \frac{2\gamma}{\beta}. \tag{7.18}$$

This fundamental result is also referred to as a *fluctuation-dissipation relation*, since it connects the noise intensity Γ of the thermal force ("fluctuations") with the friction coefficient γ ("dissipation") via the inverse temperature. We will encounter further examples of such fluctuation-dissipation relations below.

The correlation function Eq. 7.15 is interesting in its own right, since it provides information on the decay of correlations of the particle's velocity. We see that in the equilibrium regime, that is, when $t_1, t_2 \to \infty$, the decay is a simple exponential: $\left\langle V(\tau)V(0)\right\rangle_{\text{eq}} \propto e^{-\tau/\tau_r}$, where we set the time difference $|t_2 - t_1| = \tau$. The time scale of the decay is the relaxation time τ_r. We see that due to the exponential decay of the correlation function the velocity becomes essentially uncorrelated after times τ of the order of a few τ_r.

7.2.2 *Particle Position*

In order to obtain the full information about the particle dynamics, we need to consider the particle's position in addition to the velocity. The position is generally related to the velocity via $\dot{X}(t) = V(t)$. However, only in the special cases of diffusion without a force or with a constant force $-U'(X)$, the particle's velocity is uncoupled from the position, such that one can obtain the position by a simple integration of $V(t)$

$$X(t) = x_0 + \int_0^t \mathrm{d}s\, V(s), \tag{7.19}$$

for a given initial position $X(0) = x_0$. For any non-constant force $-U'(X)$, the full 2nd order Langevin Eq. 7.1 needs to be solved. This can be done, e.g., for a parabolic potential $U(x)$ as discussed in Section 7.2.3. In the following, we consider the force-free case, where $V(t)$ is described by Eq. 7.6 and the position by Eq. 7.19. Clearly, the position is also a fluctuating quantity that depends on an individual realization of $V(t)$. With Eq. 7.9 we obtain immediately for the average position

$$\left\langle X(t)\right\rangle - x_0 = \int_0^t \mathrm{d}s \left\langle V(s)\right\rangle = v_0\tau_r\left(1 - e^{-t/\tau_r}\right). \tag{7.20}$$

A measure for the diffusive transport is the mean square displacement (MSD): $\left\langle (X(t) - x_0)^2\right\rangle$. In order to calculate the MSD, we use

Eq. 7.19 to write

$$(X(t) - x_0)^2 = \int_0^t \mathrm{d}s_1 \int_0^t \mathrm{d}s_2 V(s_1)V(s_2). \tag{7.21}$$

With the velocity correlation function Eq. 7.15 we obtain thus

$$\begin{aligned}
\langle (X(t) - x_0)^2 \rangle &= \int_0^t \mathrm{d}s_1 \int_0^t \mathrm{d}s_2 \langle V(s_1)V(s_2)\rangle \\
&= \left(v_0^2 - \frac{\Gamma\tau_r}{2m^2}\right) \int_0^t \mathrm{d}s_1 \int_0^t \mathrm{d}s_2 e^{-(s_1+s_2)/\tau_r} \\
&\quad + \frac{\Gamma\tau_r}{2m^2} \int_0^t \mathrm{d}s_2 \int_0^t \mathrm{d}s_1 e^{-|s_2-s_1|/\tau_r} \\
&= \left(v_0^2 - \frac{\Gamma\tau_r}{2m^2}\right) \tau_r^2 \left(1 - e^{-t/\tau_r}\right)^2 \\
&\quad + \frac{\Gamma\tau_r}{m^2} \int_0^t \mathrm{d}s_2 \int_0^{s_2} \mathrm{d}s_1 e^{-(s_2-s_1)/\tau_r} \\
&= \left(v_0^2 - \frac{\Gamma\tau_r}{2m^2}\right) \tau_r^2 \left(1 - e^{-t/\tau_r}\right)^2 \\
&\quad + \frac{\Gamma\tau_r}{m^2} \left(\tau_r t - \tau_r^2 \left(1 - e^{-t/\tau_r}\right)\right).
\end{aligned} \tag{7.22}$$

Equation 7.22 indicates that the particle's motion is ballistic for short times t: $\langle (X(t) - x_0)^2 \rangle \propto t^2$. This can be seen by expanding e^{-t/τ_r} up to quadratic terms in t. For large times on the other hand, the dominant term in Eq. 7.22 is linear in t, which is characteristic of a diffusion process. The empirically observed diffusion coefficient D is related to the mean square displacement for large times t as

$$\langle (X(t) - x_0)^2 \rangle = 2Dt. \tag{7.23}$$

Therefore, using Eq. 7.22 we see that the diffusion coefficient can be expressed in terms of the strength of the random force as

$$D = \lim_{t\to\infty} \frac{\langle (X(t) - x_0)^2 \rangle}{2t} = \frac{\Gamma\tau_r^2}{2m^2}. \tag{7.24}$$

With Eq. 7.18, we obtain then

$$D = \frac{\tau_r}{\beta m} = \frac{1}{\gamma\beta}. \tag{7.25}$$

This is the celebrated Einstein relation, connecting a macroscopically observable quantity (the diffusion coefficient D) with the properties of the thermal environment described by β. In particular, one can determine Avogadro's constant (via Boltzmann's constant k_B) and thus infer a molecular property from a macroscopic measurement. This has been discussed for the first time in Einstein's landmark paper from 1905 (Einstein, 1956), which has provided the crucial evidence for the existence of an atomistic structure of matter.

7.2.3 *Correlations and Response*

Let us now include the effect of a potential. An exactly solvable case is that of a parabolic potential of strength κ: $U(x) = \frac{1}{2}\kappa x^2$. This means that Eq. 7.1 becomes a 2nd order linear Langevin equation for $X(t)$

$$\ddot{X}(t) + \frac{1}{\tau_r}\dot{X}(t) + \omega_0^2 X(t) = \frac{1}{m}\xi(t). \tag{7.26}$$

Here, we define the frequency $\omega_0^2 = \kappa/m$. Clearly, Eq. 7.26 is the equation of motion of a damped oscillator driven by the thermal fluctuations $\xi(t)$. An elegant way to solve this equation is to introduce $V(t) = \dot{X}(t)$ and write Eq. 7.26 in two-component form

$$\dot{X}(t) = V(t) \tag{7.27}$$

$$\dot{V}(t) = -V(t)/\tau_r - \omega_0^2 X(t) + \frac{1}{m}\xi(t) \tag{7.28}$$

We can then use vector matrix notation

$$\frac{\mathrm{d}}{\mathrm{d}t}\begin{pmatrix} X(t) \\ V(t) \end{pmatrix} = \begin{pmatrix} 0 & 1 \\ -\omega_0^2 & -\frac{1}{\tau_r} \end{pmatrix}\begin{pmatrix} X(t) \\ V(t) \end{pmatrix} + \begin{pmatrix} 0 \\ \frac{1}{m}\xi(t) \end{pmatrix} \tag{7.29}$$

or

$$\dot{\mathbf{Q}}(t) = \mathrm{M}\mathbf{Q}(t) + \boldsymbol{\xi}(t) \tag{7.30}$$

where we define

$$\mathbf{Q}(t) = \begin{pmatrix} X(t) \\ V(t) \end{pmatrix}, \ \mathrm{M} = \begin{pmatrix} 0 & 1 \\ -\omega_0^2 & -\frac{1}{\tau_r} \end{pmatrix}, \ \boldsymbol{\xi}(t) = \begin{pmatrix} 0 \\ \frac{1}{m}\xi(t) \end{pmatrix}. \tag{7.31}$$

The solution of Eq. 7.30 can be expressed in terms of the matrix exponential $e^{\mathrm{M}t}$

$$\mathbf{Q}(t) = e^{\mathrm{M}t}\mathbf{Q}_0 + e^{\mathrm{M}t}\int_0^t e^{-\mathrm{M}t'}\boldsymbol{\xi}(t')\mathrm{d}t'. \tag{7.32}$$

Let us now consider the covariance matrix

$$\left\langle \mathbf{Q}(t_2)\mathbf{Q}(t_1)^{\mathrm{T}}\right\rangle = \begin{pmatrix} \langle X(t_2)X(t_1)\rangle & \langle X(t_2)V(t_1)\rangle \\ \langle X(t_2)V(t_1)\rangle & \langle V(t_2)V(t_1)\rangle \end{pmatrix}, \tag{7.33}$$

which contains the important information on the decay of correlations of the particle's position and velocity. We obtain with Eq. 7.32

$$\begin{aligned} \left\langle \mathbf{Q}(t_2)\mathbf{Q}(t_1)^{\mathrm{T}}\right\rangle &= e^{\mathrm{M}t_2}\int_0^{t_2} \mathrm{d}t_2' \int_0^{t_1} \mathrm{d}t_1' e^{-\mathrm{M}t_2'} \\ &\times \left\langle \boldsymbol{\xi}(t_2')\boldsymbol{\xi}(t_1')\right\rangle e^{-\mathrm{M}^{\mathrm{T}}t_1'} e^{\mathrm{M}^{\mathrm{T}}t_1}, \end{aligned} \tag{7.34}$$

where one has to take care of the proper order of the transpose matrices. For simplicity, we also assume $\mathbf{Q}_0 = (0, 0)^{\mathrm{T}}$. Let us now write

$$\left\langle \boldsymbol{\xi}(t_2')\boldsymbol{\xi}(t_1')\right\rangle = \Lambda\delta(t_2' - t_1'), \qquad \Lambda = \begin{pmatrix} 0 & 0 \\ 0 & \frac{\Gamma}{m^2} \end{pmatrix}. \tag{7.35}$$

Assuming $t_2 > t_1$ then yields

$$\left\langle \mathbf{Q}(t_2)\mathbf{Q}(t_1)^{\mathrm{T}}\right\rangle = e^{\mathrm{M}t_2}\int_0^{t_1} \mathrm{d}t_1' e^{-\mathrm{M}t_1'}\Lambda e^{-\mathrm{M}^{\mathrm{T}}t_1'} e^{\mathrm{M}^{\mathrm{T}}t_1} \tag{7.36}$$

$$\begin{aligned} &= e^{\mathrm{M}(t_2-t_1)} e^{\mathrm{M}t_1} \\ &\times \int_0^{t_1} \mathrm{d}t_1' e^{-\mathrm{M}t_1'}\Lambda e^{-\mathrm{M}^{\mathrm{T}}t_1'} e^{\mathrm{M}^{\mathrm{T}}t_1} \end{aligned} \tag{7.37}$$

$$= e^{\mathrm{M}(t_2-t_1)}\left\langle \mathbf{Q}(t_1)\mathbf{Q}(t_1)^{\mathrm{T}}\right\rangle \tag{7.38}$$

Likewise for $t_1 > t_2$. In the equilibrium regime, where $t_1, t_2 \to \infty$, we have thus overall

$$\left\langle \mathbf{Q}(t_2)\mathbf{Q}(t_1)^{\mathrm{T}}\right\rangle_{\mathrm{eq}} = e^{\mathrm{M}|t_2-t_1|}\left\langle \mathbf{Q}\mathbf{Q}^{\mathrm{T}}\right\rangle_{\mathrm{eq}}, \tag{7.39}$$

which can also be written as

$$\left\langle \mathbf{Q}(\tau)\mathbf{Q}(0)^{\mathrm{T}}\right\rangle_{\mathrm{eq}} = e^{\mathrm{M}|\tau|}\left\langle \mathbf{Q}\mathbf{Q}^{\mathrm{T}}\right\rangle_{\mathrm{eq}}. \tag{7.40}$$

Furthermore, we can relate the equilibrium correlation matrix to the matrix Λ. This will provide us with an extension of Eq. 7.17 to the case of a multi-component linear system. From the result

$$\left\langle \mathbf{Q}\mathbf{Q}^{\mathrm{T}}\right\rangle_{\mathrm{eq}} = \lim_{t\to\infty} e^{\mathrm{M}t}\int_0^{t} \mathrm{d}t' e^{-\mathrm{M}t'}\Lambda e^{-\mathrm{M}^{\mathrm{T}}t'} e^{\mathrm{M}^{\mathrm{T}}t} \tag{7.41}$$

$$= \int_0^{\infty} \mathrm{d}t' e^{\mathrm{M}t'}\Lambda e^{\mathrm{M}^{\mathrm{T}}t'} \tag{7.42}$$

we derive

$$\begin{aligned}
&\mathrm{M}\left\langle \mathbf{Q}\mathbf{Q}^{\mathrm{T}}\right\rangle_{\mathrm{eq}} + \left\langle \mathbf{Q}\mathbf{Q}^{\mathrm{T}}\right\rangle_{\mathrm{eq}} \mathrm{M}^{\mathrm{T}} \\
&= \mathrm{M}\int_0^{\infty} \mathrm{d}t' e^{\mathrm{M}t'} \Lambda e^{\mathrm{M}^{\mathrm{T}}t'} + \int_0^{\infty} \mathrm{d}t' e^{\mathrm{M}t'} \Lambda e^{\mathrm{M}^{\mathrm{T}}t'} \mathrm{M}^{\mathrm{T}} \\
&= \int_0^{\infty} \mathrm{d}t' \frac{\mathrm{d}}{\mathrm{d}t'} e^{\mathrm{M}t'} \Lambda e^{\mathrm{M}^{\mathrm{T}}t'} \\
&= -\Lambda.
\end{aligned} \tag{7.43}$$

In the last step we have used the fact that the matrix M has only negative real eigenvalues and thus the contribution at $t = \infty$ vanishes. This leads to

$$\mathrm{M}\left\langle \mathbf{Q}\mathbf{Q}^{\mathrm{T}}\right\rangle_{\mathrm{eq}} + \left\langle \mathbf{Q}\mathbf{Q}^{\mathrm{T}}\right\rangle_{\mathrm{eq}} \mathrm{M}^{\mathrm{T}} = -\Lambda. \tag{7.44}$$

Note that from the equipartition theorem we know that

$$\left\langle E_{\mathrm{pot}}\right\rangle_{\mathrm{eq}} = \frac{1}{2}\kappa\left\langle X^2\right\rangle_{\mathrm{eq}} = \frac{1}{2\beta} \tag{7.45}$$

$$\left\langle E_{\mathrm{kin}}\right\rangle_{\mathrm{eq}} = \frac{1}{2}m\left\langle V^2\right\rangle_{\mathrm{eq}} = \frac{1}{2\beta} \tag{7.46}$$

Therefore, $\left\langle \mathbf{Q}\mathbf{Q}^{\mathrm{T}}\right\rangle_{\mathrm{eq}}$ has the entries

$$\left\langle \mathbf{Q}\mathbf{Q}^{\mathrm{T}}\right\rangle_{\mathrm{eq}} = \begin{pmatrix} \left\langle X^2\right\rangle_{\mathrm{eq}} & 0 \\ 0 & \left\langle V^2\right\rangle_{\mathrm{eq}} \end{pmatrix} = \begin{pmatrix} \frac{1}{\kappa\beta} & 0 \\ 0 & \frac{1}{m\beta} \end{pmatrix} \tag{7.47}$$

and the cross terms $\langle XV\rangle_{\mathrm{eq}} = \langle X\rangle_{\mathrm{eq}}\langle V\rangle_{\mathrm{eq}} = 0$. Therefore, the statement of Eq. 7.44 is equivalent to Eq. 7.18: Eq. 7.44 connects the noise intensity Λ with the matrix M containing the deterministic dynamics. Equation 7.44 is also referred to as a fluctuation-dissipation relation.

Instead of Eq. 7.39, we can derive an alternative form of the correlation matrix $\left\langle \mathbf{Q}(\tau)\mathbf{Q}(0)^{\mathrm{T}}\right\rangle_{\mathrm{eq}}$ in Fourier space. To this end, let us introduce the power spectral density, which is just the Fourier transform of $\langle \mathbf{Q}(\tau)\mathbf{Q}(0)^{\mathrm{T}} \rangle_{\text{eq}}$:

$$S(\omega) = \int_{-\infty}^{\infty} \mathrm{d}\omega \left\langle \mathbf{Q}(\tau)\mathbf{Q}(0)^{\mathrm{T}} \right\rangle_{\text{eq}} e^{i\omega\tau} \tag{7.48}$$

$$= \int_{-\infty}^{\infty} \mathrm{d}\omega e^{\mathrm{M}|\tau|+i\omega\tau} \left\langle \mathbf{Q}\mathbf{Q}^{\mathrm{T}} \right\rangle_{\text{eq}} \tag{7.49}$$

$$= \int_{0}^{\infty} \mathrm{d}\omega e^{\mathrm{M}\tau+i\omega\tau} \left\langle \mathbf{Q}\mathbf{Q}^{\mathrm{T}} \right\rangle_{\text{eq}} + \left\langle \mathbf{Q}\mathbf{Q}^{\mathrm{T}} \right\rangle_{\text{eq}} \int_{-\infty}^{0} \mathrm{d}\omega e^{-\mathrm{M}^{\mathrm{T}}\tau+i\omega\tau} \tag{7.50}$$

$$= -\left(\frac{1}{\mathrm{M}+i\omega} \left\langle \mathbf{Q}\mathbf{Q}^{\mathrm{T}} \right\rangle_{\text{eq}} + \left\langle \mathbf{Q}\mathbf{Q}^{\mathrm{T}} \right\rangle_{\text{eq}} \frac{1}{\mathrm{M}^{\mathrm{T}}-i\omega} \right). \tag{7.51}$$

This means that

$$(\mathrm{M}+i\omega)S(\omega)(\mathrm{M}^{\mathrm{T}}-i\omega) = -\left\langle \mathbf{Q}\mathbf{Q}^{\mathrm{T}} \right\rangle_{\text{eq}} (\mathrm{M}^{\mathrm{T}}-i\omega) - (\mathrm{M}+i\omega) \left\langle \mathbf{Q}\mathbf{Q}^{\mathrm{T}} \right\rangle_{\text{eq}} \tag{7.52}$$

$$= -\left\langle \mathbf{Q}\mathbf{Q}^{\mathrm{T}} \right\rangle_{\text{eq}} \mathrm{M}^{\mathrm{T}} - \mathrm{M} \left\langle \mathbf{Q}\mathbf{Q}^{\mathrm{T}} \right\rangle_{\text{eq}} \tag{7.53}$$

$$= \Lambda \tag{7.54}$$

using Eq. 7.44. Therefore

$$S(\omega) = \frac{1}{\mathrm{M}+i\omega} \Lambda \frac{1}{\mathrm{M}^{\mathrm{T}}-i\omega}. \tag{7.55}$$

This means that we can obtain the correlation matrix $\langle \mathbf{Q}(\tau)\mathbf{Q}(0)^{\mathrm{T}} \rangle_{\text{eq}}$ either by evaluating the matrix exponential in Eq. 7.39 together with Eq. 7.47, or by evaluating the product of matrices in Eq. 7.55 and then performing an inverse Fourier transform. It is quite common to obtain the power spectrum directly in an experiment, in which case the representation Eq. 7.55 is more useful.

7.2.3.1 Response functions

When a system at equilibrium is exposed to a time-dependent perturbation it will be driven away from equilibrium during the duration of the perturbation. After the perturbation has stopped, the system will eventually settle back in its equilibrium state after some time period. In the picture of our model Eq. 7.1, we can imagine the perturbation to be an additional force acting on the particle, say $h(t)$, with a certain magnitude and time dependence. However, since the particle is subject to the random thermal force $\xi(t)$ it is not able to distinguish, whether $h(t)$ is indeed imposed by some external agent, or whether $h(t)$ simply represents a large untypical fluctuation exerted by the thermal force. As a consequence, we expect that the response of the system to a perturbation will contain information of the equilibrium state. As we will see below, the response is in fact related to the correlation function of the observable that is conjugate to the perturbation, i.e., the particle position. On a quantitative level, we can express the response to a time-dependent perturbation $h(t)$ in a linear approximation as

$$y(t) = \int_{-\infty}^{t} \mathrm{d}t' \chi(t-t')h(t'). \tag{7.56}$$

Here, $\chi(t)$ is the *linear response function* and $y(t)$ the observable conjugate to the perturbation (the position). Due to causality, the response of the system to a perturbation at $t' > t$ is zero, therefore we can also write

$$y(t) = \int_{-\infty}^{\infty} \mathrm{d}t' \chi(t-t')h(t'). \tag{7.57}$$

In this way we can use a Fourier transform to obtain the response function as

$$\tilde{\chi}(\omega) = \frac{\tilde{y}(\omega)}{\tilde{h}(\omega)}, \tag{7.58}$$

where the $\sim$ indicates the Fourier transformed functions.

Let us look more closely at the harmonic oscillator driven by a thermal force described by Eq. 7.26. A Fourier transformation of Eq. 7.26 yields

$$-\omega^2 \tilde{X}(\omega) - \frac{i\omega}{\tau_r}\tilde{X}(\omega) + \omega_0^2 \tilde{X}(\omega) = \frac{1}{m}\tilde{\xi}(\omega). \tag{7.59}$$

If we treat the random force as a time-dependent perturbation we can set $\tilde{h}(\omega) = \tilde{\xi}(\omega)$, such that the response function Eq. 7.58 is given by

$$\tilde{\chi}(\omega) = \frac{1}{m(\omega_0^2 - \omega^2 - i\omega/\tau_r)}, \tag{7.60}$$

which has the imaginary part

$$\mathrm{Im}\{\tilde{\chi}(\omega)\} = \frac{\omega}{m\tau_r((\omega_0^2 - \omega^2)^2 + \omega^2/\tau_r^2)}. \tag{7.61}$$

We compare this result with the Fourier transform of the position correlation function $S_{xx}(\omega) = \int_{-\infty}^{\infty} \mathrm{d}\tau \left\langle X(\tau)X(0)\right\rangle_{\mathrm{eq}} e^{i\omega\tau}$, which is obtained from Eq. 7.55 by evaluating the matrix products. This leads to

$$S_{xx}(\omega) = \frac{\Gamma}{m^2((\omega_0^2 - \omega^2)^2 + \omega^2/\tau_r^2)}. \tag{7.62}$$

Combining Eqs. 7.61 and 7.62 yields

$$\mathrm{Im}\{\tilde{\chi}(\omega)\} = \frac{m\omega}{\Gamma\tau_r} S_{xx}(\omega) \tag{7.63}$$

Using Eq. 7.18, we obtain the relationship between the response function and the position correlation function in Fourier space:

$$\mathrm{Im}\{\tilde{\chi}(\omega)\} = \frac{\omega\beta}{2} S_{xx}(\omega). \tag{7.64}$$

Equation 7.64 is the generic form of a fluctuation-dissipation relation involving response functions. Even though it is here derived for the special case of a driven oscillator, Eq. 7.64 can be shown to hold under very general conditions (Chaikin and Lubensky, 2000).

7.2.4 *Simulation of Langevin Equations*

It is often difficult if not impossible to derive analytical expressions for the moments of $X(t)$ or the full pdf $f_X(x, t)$ given a general potential U. Solving the equation of motion Eq. 7.1 numerically is thus often the only way to obtain quantitative theoretical predictions. For generality, we consider a generic variable $Z(t)$ (velocity or position), whose time evolution is described by the Langevin equation

$$\dot{Z}(t) = F(Z(t)) + \xi(t), \tag{7.65}$$

where $F(z)$ denotes an arbitrary force and $\xi(t)$ is a random force with the properties Eqs. 7.8–7.10 as before. As initial condition we assume $Z(0) = z_0$. Starting point for a simulation is a suitable time discretization of the Langevin Eq. 7.65, which we can formulate analogous to conventional discretization schemes for ordinary differential equations, e.g., the Euler scheme. The only difference is the discretization of the random force $\xi(t)$, which requires special consideration. Introducing a time step Δt we write Eq. 7.65 in discretized form for $Z_n = Z(n\Delta t)$

$$Z_{n+1} - Z_n = F(Z_n)\Delta t + \xi_n. \tag{7.66}$$

We define the increments ξ_n of the random force $\xi(t)$ as

$$\xi_n = \int_{t_n}^{t_n+\Delta t} \mathrm{d}t' \xi(t'), \tag{7.67}$$

where $t_n = n\Delta t$. Since $\xi(t)$ is a Gaussian random variable, also the ξ_n are Gaussian. We see that the average $\langle \xi_n \rangle = 0$ due to Eq. 7.8 and the variance $\mathrm{Var}(\xi_n)$ is

$$\begin{aligned} \mathrm{Var}(\xi_n) &= \langle \xi_n^2 \rangle - \langle \xi_n \rangle^2 \\ &= \langle \xi_n^2 \rangle \\ &= \int_{t_n}^{t_n+\Delta t} \mathrm{d}t' \int_{t_n}^{t_n+\Delta t} \mathrm{d}t'' \langle \xi(t')\xi(t'') \rangle \\ &= \Gamma \Delta t \end{aligned} \tag{7.68}$$

due to Eq. 7.10. Likewise, one obtains that $\langle \xi_i \xi_j \rangle = \Gamma \Delta t \delta_{ij}$, since increments over different non-overlapping time periods are statistically independent. Note that δ_{ij} denotes the Kronecker delta. In the next step, we need the following general property of a Gaussian random variable: If Y is a Gaussian random variable with average μ and variance σ^2, then Y can be written as $Y = \mu + \sigma W$, where W is a Gaussian random variable with a zero average and variance 1. Since $\langle \xi_n \rangle = 0$ and $\mathrm{Var}(\xi_n) = \Gamma \Delta t$, we can thus write

$$\xi_n = \sqrt{\Gamma \Delta t}\, \zeta_n, \tag{7.69}$$

where ζ_n is a Gaussian random variable with a zero average and variance 1. Eq. 7.66 then becomes

$$Z_{n+1} - Z_n = F(Z_n)\Delta t + \sqrt{\Gamma \Delta t}\, \zeta_n. \tag{7.70}$$

All the ζ_n are identically and independently distributed and are conveniently generated by standard algorithms for Gaussian random variables such as the Box-Müller algorithm. By iterating Eq. 7.70 starting at z_0 a trajectory of $Z(s) : 0 \leqslant s \leqslant t$ is generated (see Fig. 7.1a).

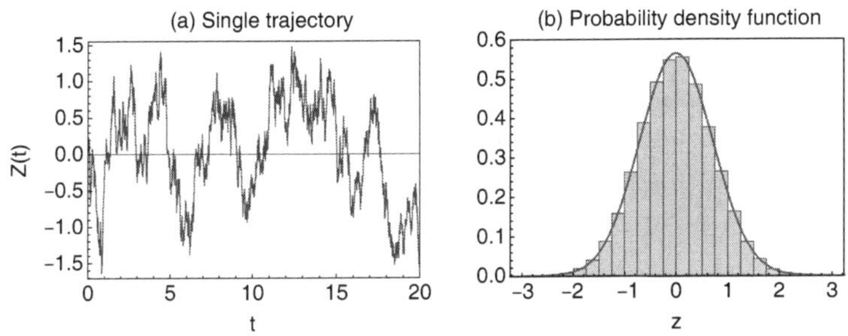

Figure 7.1 (a) A stochastic trajectory generated by Eq. 7.70 for the case of a linear force: $F(z) = -z$ (Ornstein-Uhlenbeck process). (b) The pdf is obtained from a normalized histogram of the values $Z(t)$ following the steps 1–4 in Section 7.2.4.2.

In order to calculate averages of $Z(t)$, we require a large number N of realizations of $Z(t)$ all generated with the same initial condition z_0. The average Eq. 7.3 is then formally $\langle g(Z(t)) \rangle = \lim_{N\to\infty} \hat{g}_N$, where we define

$$\hat{g}_N = \frac{1}{N} \sum_{i=1}^{N} g(Z_{k,i}). \tag{7.71}$$

The value $Z_{k,i}$ is the discretized value Z_k at the time $t = k\Delta t$ of the ith trajectory. For any finite N, $\hat{g}_N$ is only an approximation of the true average $\langle g(Z(t)) \rangle$ that is sought after. Clearly, we would like to have an estimation of the error of the approximation $\hat{g}_N$ for a given N. We discuss this briefly in the following.

7.2.4.1 Error estimation

The estimate Eq. 7.71, also called *Monte-Carlo estimate*, is itself a random variable, since it consists of a finite sum of random variables. Therefore, we can consider the mean and variance of this random variable:

$$\langle \hat{g}_N \rangle = \frac{1}{N} \sum_{i=1}^{N} \langle g(Z_{k,i}) \rangle = \langle g(Z_k) \rangle = \langle g(Z(t)) \rangle. \tag{7.72}$$

The second equality follows from the fact that the values $Z_{k,i}$ are all identically distributed for all i, so that one does not have to distinguish between them and can drop the i index. The last equality follows if we assume that Δt is sufficiently small, so that Z_k represents a realization of the true $Z(t)$. Equation 7.72 indicates that $\hat{g}_N$ is an unbiased estimate, i.e., its average is precisely the ensemble average that one wants to determine, namely $\langle g(Z(t)) \rangle$. The variance of $\hat{g}_N$ is

$$\begin{aligned} \mathrm{Var}(\hat{g}_N) &= \frac{1}{N^2} \mathrm{Var}\left(\sum_{i=1}^{N} g(Z_{k,i}) \right) \\ &= \frac{1}{N^2} \sum_{i=1}^{N} \mathrm{Var}(g(Z_{k,i})). \end{aligned} \tag{7.73}$$

The last step follows because the values $Z_{k,i}$ are all statistically independent from each other. Since they are also all identically distributed one can drop the i index again and obtain

$$\mathrm{Var}(\hat{g}_N) = \frac{1}{N^2} \sum_{i=1}^{N} \mathrm{Var}(g(Z_k)) = \frac{1}{N} \mathrm{Var}(g(Z(t))) \to 0 \tag{7.74}$$

for $N \to \infty$. We see that the variance converges to zero for large N, so that the Monte-Carlo estimate converges indeed to the true average in this limit. As a measure of the error one typically considers $\sqrt{\mathrm{Var}(\hat{g}_N)}$. However, since $\mathrm{Var}(\hat{g}_N) = \left\langle (\hat{g}_N - \langle \hat{g}_N \rangle)^2 \right\rangle$ and $\langle \hat{g}_N \rangle$ is just the average that is sought after, we can not determine the variance in the simulation. The solution is to estimate $\mathrm{Var}(g(Z(t)))$ as a sample variance from the simulation values $Z_{k,i}$. This means that we determine the quantity

$$\widehat{\mathrm{Var}}(g(Z(t))) = \frac{1}{N-1} \sum_{i=1}^{N} \left(g(Z_{k,i}) - \hat{g}_N \right)^2 \tag{7.75}$$

as an estimate of $\mathrm{Var}(g(Z(t)))$. We obtain then for $\mathrm{Var}(\hat{g}_N)$ with Eq. 7.74

$$\begin{aligned} \mathrm{Var}(\hat{g}_N) &= \frac{1}{N} \mathrm{Var}(g(Z(t))) \\ &\approx \frac{1}{N} \widehat{\mathrm{Var}}(g(Z(t))) \\ &= \frac{1}{N(N-1)} \sum_{i=1}^{N} \left(g(Z_{k,i}) - \hat{g}_N \right)^2 . \end{aligned} \tag{7.76}$$

The error of the Monte-Carlo estimate is thus

$$\varepsilon = \sqrt{\mathrm{Var}(\hat{g}_N)} \approx \sqrt{\frac{1}{N(N-1)} \sum_{i=1}^{N} (g(Z_{k,i}) - \hat{g}_N)^2}, \quad (7.77)$$

which is also denoted as *root-mean-square error* ε. Since Var $(g(Z(t)))$ is independent of N we see that ε scales as

$$\varepsilon = \sqrt{\frac{1}{N} \mathrm{Var}(g(Z(t)))} \propto \frac{1}{\sqrt{N}}. \quad (7.78)$$

This means that in order to reduce the error, e.g., by a factor 10, we need to increase the number of samples N by a factor 100.

7.2.4.2 Determining the probability density function

From the simulation data we obtain an approximation of the probability density function by constructing a histogram of the values $Z(t)$. For a fixed initial condition z_0 the pdf will depend explicitly on z_0. Therefore, we indicate the initial condition in the pdf as $f_Z(z, t|z_0)$, which is more precisely the transition probability of the process: $f_z(z, t|z_0)\mathrm{d}z$ is the probability to observe a value of $Z(t)$ in the interval $[z, z + \mathrm{d}z]$ at time t, given the process starts from z_0 at time 0. We obtain the transition probability numerically as follows:

(1) Generate N realizations of the random variable $Z(t)$ at time $t = k\Delta t$: $\{Z_{k,1}, Z_{k,2}, \ldots, Z_{k,N}\}$. Every iteration starts at z_0 at time 0. The values $Z_{k,i}$ fall into a range $[a, b]$.

(2) A histogram can be constructed with a suitable discretization of the range of Z values into bins of width $\Delta z = \frac{b-a}{m}$, assuming m bins. The bins correspond to intervals $[z_j, z_j + \Delta z]$, with $z_j = a + (j-1)\Delta z$ and $j \in [1, m]$. Counting the number of times a value $Z_{k,i}$ falls into a bin $[z_j, z_j + \Delta z]$ yields the histogram value h_j. Clearly, we need to have $\sum_{j=1}^{m} h_j = N$, since any value $Z_{k,i}$ has to fall into one of the m bins.

(3) The probability to observe a value of $Z(t)$ in the jth interval $[z_j, z_j + \Delta z]$ corresponds to the relative frequency

$$\Pr(z_j < Z(t) \leqslant z_j + \Delta z) = \frac{h_j}{N}. \quad (7.79)$$

(4) The exact pdf is obtained in the limit in which both the discretization $\Delta z \to \infty$ (corresponding to an infinite number of bins m in the interval $[a, b]$) and the number of realizations $N \to \infty$

$$f_Z(z_j, t|z_0) = \lim_{N\to\infty, \Delta z\to 0} \frac{\Pr(z_j < Z(t) \leqslant z_j + \Delta z)}{\Delta z}. \quad (7.80)$$

For every finite value of N and m the pdf obtained in this way is an approximation of the exact pdf (see Fig. 7.1b).

7.3 Probability Density Functions

In order to calculate the pdf $f_Z(z, t|z_0)$ analytically we derive from the Langevin Eq. 7.65 a partial differential equation for the pdf known as *Fokker-Planck equation*. It is important to note that while the Langevin equation is a stochastic equation, i.e., its solutions are random trajectories that are different in each realization, the Fokker-Planck equation is deterministic: Its solution contains the probability that the trajectory reaches a certain point at a given time. In order to derive the Fokker-Planck equation, we first discuss a mathematical result known as Itô's formula. This formula extends the familiar chain rule from ordinary calculus to stochastic processes, forming the basis of stochastic calculus. In the following, we provide a heuristic derivation of Itô's formula and discuss the derivation of the Fokker-Planck equation.

7.3.1 *Itô's Formula and the Fokker-Planck Equation*

We keep the general formulation of a diffusion process, Eq. 7.65, and start with the increment of the process $Z(t)$ defined as

$$\Delta Z = Z(t + \Delta t) - Z(t). \quad (7.81)$$

A Taylor expansion of a function $g(Z(t))$ up to quadratic orders in ΔZ yields

$$g(Z(t + \Delta t)) = g(Z(t)) + g'(Z(t))\Delta Z + \frac{1}{2}g''(Z(t))\Delta Z^2, \quad (7.82)$$

where we write $g'(x) = \mathrm{d}g(x)/\mathrm{d}x$. We then replace ΔZ by its discretized version Eq. 7.70

$$\begin{aligned}\Delta Z &= Z(t+\Delta t) - Z(t) \\ &= Z_{n+1} - Z_n = F(Z_n)\Delta t + \sqrt{\Gamma \Delta t}\,\zeta_n. \end{aligned} \tag{7.83}$$

Therefore

$$\begin{aligned}(\Delta Z)^2 &= \Gamma \Delta t\, \zeta_n^2 + \sqrt{\Gamma \Delta t}\,\zeta_n F(Z_n)\Delta t + (F(Z_n)\Delta t)^2 \\ &\approx \Gamma \Delta t + \mathcal{O}(\Delta t^{3/2}). \end{aligned} \tag{7.84}$$

In the last step, we have approximated ζ_n^2 by its expected value $\langle \zeta_n^2 \rangle = 1$, which shows that the dominant term in $(\Delta Z)^2$ is linear in Δt. Substituting ΔZ and $(\Delta Z)^2$ from Eq. 7.84 into Eq. 7.82 and keeping only terms linear in Δt and ΔZ, yields

$$g(Z(t+\Delta t)) = g(Z(t)) + g'(Z(t))\Delta Z + \frac{\Gamma}{2} g''(Z(t))\Delta t. \tag{7.85}$$

Dividing by Δt and taking the limit of $\Delta t \to 0$, we can also write

$$\frac{\mathrm{d}}{\mathrm{d}t} g(Z(t)) = g'(Z(t))\dot{Z}(t) + \frac{\Gamma}{2} g''(Z(t)). \tag{7.86}$$

We see that compared with the usual chain rule, the additional term $\frac{\Gamma}{2} g''(Z(t))$ shows up due to the fact $Z(t)$ is not a smooth function, but given by a diffusion process, which is characterized by very "rough" variations (see Fig. 7.1a). Equation 7.86 is commonly referred to as *Itô's formula*.

We can use this result to derive an evolution equation for the pdf conditional on an initial value $Z(0) = z_0$. We substitute Eq. 7.65 for $\dot{Z}(t)$ in Eq. 7.86 and take the expected value on both sides. This yields

$$\begin{aligned}\left\langle \frac{\mathrm{d}}{\mathrm{d}t} g(Z(t)) \right\rangle &= \langle g'(Z(t))(F(Z(t)) + \xi(t)) \rangle \\ &\quad + \frac{\Gamma}{2} \langle g''(Z(t)) \rangle . \end{aligned} \tag{7.87}$$

We can then express the average explicitly in terms of the pdf $f_Z(z, t|z_0)$ using Eq. 7.4 (the average is strictly an average conditional on the initial value z_0). On the lhs we obtain

$$\left\langle \frac{\mathrm{d}}{\mathrm{d}t} g(Z(t)) \right\rangle = \frac{\partial}{\partial t} \int_{-\infty}^{\infty} \mathrm{d}z\, g(z) f_Z(z, t|z_0) = \int_{-\infty}^{\infty} \mathrm{d}z\, g(z) \frac{\partial}{\partial t} f_Z(z, t|z_0). \tag{7.88}$$

On the rhs we have

$$\langle g'(Z(t))F(Z(t))\rangle = \int_{-\infty}^{\infty} \mathrm{d}z\, g'(z)F(z) f_Z(z, t|z_0) = -\int_{-\infty}^{\infty} \mathrm{d}z\, g(z)\frac{\partial}{\partial z}(F(z) f_Z(z, t|z_0)) \tag{7.89}$$

and

$$\frac{\Gamma}{2}\langle g''(Z(t))\rangle = \frac{\Gamma}{2}\int_{-\infty}^{\infty} \mathrm{d}z\, g''(z) f_Z(z, t|z_0) = \frac{\Gamma}{2}\int_{-\infty}^{\infty} \mathrm{d}z\, g(z)\frac{\partial^2}{\partial z^2} f_Z(z, t|z_0). \tag{7.90}$$

In the last steps we have performed partial integrations and used the fact that the boundary terms vanish: In order to be normalizable the pdf needs to satisfy $\lim_{z\to\pm\infty} f_Z(z, t|z_0) = 0$. In addition, we require also $\lim_{z\to\pm\infty} \frac{\partial}{\partial z} f_Z(z, t|z_0) = 0$. The remaining term $\langle g'(Z(t))\xi(t)\rangle$ disappears since

$$\langle g'(Z(t))\xi(t)\rangle \approx \langle g'(Z_n)\xi_n\rangle = \langle g'(Z_n)\rangle \langle \xi_n\rangle = 0. \tag{7.91}$$

The reason for the disappearance is that in our discretization Z_n depends on all the ξ_i with $i \leqslant n-1$ only, such that Z_n and ξ_n are statistically independent. As a consequence, the average in Eq. 7.91 factorizes and the result follows from the property $\langle \xi_n\rangle = 0$. Combining Eqs. 7.88–7.90 yields

$$\int_{-\infty}^{\infty} \mathrm{d}z\, g(z)\frac{\partial}{\partial t} f_Z(z, t|z_0) = -\int_{-\infty}^{\infty} \mathrm{d}z\, g(z)\frac{\partial}{\partial z}(F(z) f_Z(z, t|z_0)) + \frac{\Gamma}{2}\int_{-\infty}^{\infty} \mathrm{d}z\, g(z)\frac{\partial^2}{\partial z^2} f_Z(z, t|z_0). \tag{7.92}$$

Since g is an arbitrary function, we can conclude that

$$\frac{\partial}{\partial t} f_Z(z, t|z_0) = -\frac{\partial}{\partial z}(F(z) f_Z(z, t|z_0)) + \frac{\Gamma}{2}\frac{\partial^2}{\partial z^2} f_Z(z, t|z_0), \quad (7.93)$$

which is the partial differential equation (Fokker-Planck equation) that we sought after. Equation 7.93 is accompanied by the initial condition $f_Z(z, 0|z_0) = \delta(z - z_0)$ indicating that the process starts at the fixed point z_0, so that the pdf at time zero is a delta function. Equation 7.93 can incorporate different boundary conditions in order to model specific scenarios that constrain the process, for example reflecting or absorbing boundaries. In the following, we always assume an infinite domain with the natural boundary conditions: $\lim_{z\to\pm\infty} f_Z(z, t|z_0) = 0$ and $\lim_{z\to\pm\infty} \frac{\partial}{\partial z} f_Z(z, t|z_0) = 0$.

7.3.2 *Ornstein-Uhlenbeck Process*

Let us revisit the case of a free diffusion process, $U(x) = 0$, such that $V(t)$ follows the Ornstein-Uhlenbeck process Eq. 7.5. By solving the Fokker-Planck equation associated with the Langevin Eq. 7.5, we can calculate the pdf $f_V(v, t|v_0, t_0)$ for the initial velocity $V(t_0) = v_0$. Note that we specify here the initial time t_0 explicitly, while previously we always set $t_0 = 0$. Basic equilibrium statistical mechanics requires that $f_V(v, t|v_0, t_0)$ reproduces the Maxwell-Boltzmann velocity distribution, when the particle is at equilibrium. Therefore, the stationary (equilibrium) limit needs to be: $\lim_{t\to\infty} f_V(v, t|v_0, t_0) = f_{eq}(v)$, where $f_{eq}(v)$ denotes the Maxwell-Boltzmann velocity distribution

$$f_{eq}(v) = \sqrt{\frac{m\beta}{2\pi}} e^{-\frac{1}{2}m\beta v^2}. \quad (7.94)$$

Let us verify this explicitly. We first write Eq. 7.5 in the form of Eq. 7.65

$$\begin{aligned} \dot{V}(t) &= -\frac{\gamma}{m} V(t) + \frac{1}{m}\xi(t) \\ &= -\frac{1}{\tau_r} V(t) + \tilde{\xi}(t), \end{aligned} \quad (7.95)$$

where $\langle \tilde{\xi}(t) \rangle = 0$ and $\langle \tilde{\xi}(t)\tilde{\xi}(t') \rangle = \frac{\Gamma}{m^2}\delta(t - t')$. Note that we write $\tilde{\xi}(t)$ because it is strictly no longer a force but has dimensions of velocity. The Fokker-Planck equation associated with Eq. 7.95 is

$$\frac{\partial}{\partial t} f_V(v, t|v_0, t_0) = \frac{\partial}{\partial v}\frac{v}{\tau_r} f_V(v, t|v_0, t_0) + \frac{\Gamma}{2m^2}\frac{\partial^2}{\partial v^2} f_V(v, t|v_0, t_0), \tag{7.96}$$

with an initial condition $f_V(v, t_0|v_0, t_0) = \delta(v - v_0)$. This equation can be solved, e.g., by using a Fourier transform and applying the method of characteristics. We do not discuss this calculation at this point, but refer the reader, e.g., to Ref. (Gardiner, 2010). The full time-dependent solution is

$$f_V(v, t|v_0, t_0) = \sqrt{\frac{m^2}{\pi\Gamma(1 - e^{-2(t-t_0)/\tau_r})\tau_r}} \times \exp\left\{-\frac{m^2\left(v - e^{-(t-t_0)/\tau_r} v_0\right)^2}{\Gamma\tau_r(1 - e^{-2(t-t_0)/\tau_r})}\right\}. \tag{7.97}$$

We see that in the stationary limit, we have

$$\lim_{t\to\infty} f_V(v, t|v_0, t_0) = \sqrt{\frac{m^2}{\pi\Gamma\tau_r}} e^{-\frac{m^2}{\Gamma\tau_r}v^2}. \tag{7.98}$$

Using the fluctuation-dissipation relation Eq. 7.18, we have $\Gamma\tau_r = 2m/\beta$ and thus $\lim_{t\to\infty} f_V(v, t|v_0, t_0)$ just reproduces the Maxwell-Boltzmann distribution Eq. 7.94.

The solution Eq. 7.97 contains also the time-dependent information of the process $V(t)$. The moments can be calculated by straightforward integration: $\langle V^m(t)\rangle = \int_{-\infty}^{\infty} \mathrm{d}v\, v^m f_V(v, t|v_0)$. For the calculation of the correlation function, we note that $\langle V(t_1)V(t_2)\rangle$ depends on two time points, t_1 and t_2, so that $f_V(v, t|v_0, t_0)$ is not sufficient by itself. More precisely, the correlation function is obtained by the double integral

$$\langle V(t_1)V(t_2)\rangle = \int_{-\infty}^{\infty} \mathrm{d}v_1 \int_{-\infty}^{\infty} \mathrm{d}v_2\, v_1 v_2 f_V(v_1, t_1; v_2, t_2|v_0, t_0), \tag{7.99}$$

where $f_V(v_1, t_1; v_2, t_2|v_0, t_0)\mathrm{d}v_1\mathrm{d}v_2$ contains the probability to observe $V(t)$ in the interval $[v_1, v_1 + \mathrm{d}v_1]$ at the time t_1 *and* in the interval $[v_2, v_2 + \mathrm{d}v_2]$ at the time t_2, given the initial condition

$V(t_0) = v_0$. This 2-point probability density function is given as a product

$$f_V(v_1, t_1; v_2, t_2|v_0, t_0) = f_V(v_2, t_2|v_1, t_1) f_V(v_1, t_1|v_0, t_0) \tag{7.100}$$

of solutions Eq. 7.97. The factorization Eq. 7.100 is valid here, because the probability to observe $V(t)$ to be at some point v_2 at the time t_2 given the process was observed at an earlier time t_1 at v_1 is fully specified by the transition probability $f_V(v_2, t_2|v_1, t_1)$, i.e., only depends on v_1, v_2 and t_1, t_2. There is no dependence, e.g., on times before t_1. The factorization property is characteristic for Markov processes and is essentially due to the fact that the Langevin equation as well as the noise do not contain any memory terms, so that the future behaviour is only determined by the present, and not by the past (Van Kampen, 2007). As an exercise, one can readily check that Eq. 7.99 reproduces the result for the velocity correlation function Eq. 7.15 obtained by a direct integration.

7.3.3 *The Multivariate Case*

In the general case of Eq. 7.1 we are interested in the properties of the position as well as the velocity. In this case, we need to consider the multivariate pdf $f_X(x, v, t|x_0, v_0, t_0)$, which contains the probability to observe $X(t)$ in the interval $[x, x+\mathrm{d}x]$ at the time t and $\dot{X}(t)$ in the interval $[v, v+\mathrm{d}v]$ also at the time t, given the initial conditions: $V(t_0) = v_0$ and $X(t_0) = x_0$. We have seen that Eq. 7.1 is more conveniently discussed in a two-component form (compare with Eqs. 7.27 and 7.28).

$$\dot{X}(t) = V(t) \tag{7.101}$$

$$\dot{V}(t) = -\frac{\gamma}{m}V(t) - \frac{1}{m}U'(X(t)) + \frac{1}{m}\xi(t). \tag{7.102}$$

Therefore, we consider a general vector $\mathbf{Z}(t) = (Z_1(t), \ldots, Z_n(t))^{\mathrm{T}}$, for which the components satisfy

$$\dot{Z}_i(t) = F_i(\mathbf{Z}(t)) + \xi_i(t). \tag{7.103}$$

The random forces $\xi_i(t)$ are specified as

$$\langle \xi_i(t) \rangle = 0, \quad \langle \xi_i(t)\xi_j(t') \rangle = \Lambda_{ij}\delta(t-t'), \quad \Lambda_{ij} = \Lambda_{ji}. \tag{7.104}$$

The following multivariate Itô formula can then be derived (Gardiner, 2010):

$$\frac{\mathrm{d}}{\mathrm{d}t}g(\mathbf{Z}(t)) = \sum_{i=1}^{n} g'(Z_i(t))\dot{Z}_i(t) + \frac{1}{2}\sum_{i,j=1}^{n} \Lambda_{ij}\frac{\partial}{\partial Z_i(t)}\frac{\partial}{\partial Z_j(t)}g(\mathbf{Z}(t)). \qquad (7.105)$$

The multivariate Fokker-Planck equation for the pdf $f_{\mathbf{Z}}(\mathbf{z}, t|\mathbf{z}_0, t_0)$ follows in analogy to Eq. 7.93

$$\frac{\partial}{\partial t}f_{\mathbf{Z}}(\mathbf{z}, t|\mathbf{z}_0, t_0) = -\sum_{i=1}^{n}\frac{\partial}{\partial z_i}(F_i(\mathbf{z})f_{\mathbf{Z}}(\mathbf{z}, t|\mathbf{z}_0, t_0)) + \frac{1}{2}\sum_{i,j=1}^{n}\Lambda_{ij}\frac{\partial}{\partial z_i}\frac{\partial}{\partial z_j}f_{\mathbf{Z}}(\mathbf{z}, t|\mathbf{z}_0, t_0), \qquad (7.106)$$

with $f(\mathbf{z}, t_0|\mathbf{z}_0, t_0) = \delta(\mathbf{z} - \mathbf{z}_0)$.

Applying Eq. 7.106 to the general case Eqs. 7.101 and 7.102, where $\mathbf{F}(x, v) = \begin{pmatrix} v \\ -(\gamma v + U'(x))/m \end{pmatrix}$ and $\Lambda = \begin{pmatrix} 0 & 0 \\ 0 & \Gamma/m^2 \end{pmatrix}$, thus yields the Fokker-Planck equation

$$\frac{\partial}{\partial t}f_X(x, v, t|x_0, v_0, t_0) = -v\frac{\partial}{\partial x}f_X(x, v, t|x_0, v_0, t_0) + \frac{1}{m}\frac{\partial}{\partial v}(\gamma v + U'(x))f_X(x, v, t|x_0, v_0, t_0) + \frac{\Gamma}{2m^2}\frac{\partial^2}{\partial v^2}f_X(x, v, t|x_0, v_0, t_0) \qquad (7.107)$$

with the initial condition $f_X(x, v, t_0|x_0, v_0, t_0) = \delta(x - x_0)\delta(v - v_0)$. The evolution Eq. 7.107 for the joint position-velocity pdf is also called *Klein-Kramers equation*. If we are only interested in the properties of the position, we need to calculate the marginal pdf by integrating out the velocity degree of freedom

$$f_X(x, t|x_0, t_0) = \int_{-\infty}^{\infty} \mathrm{d}v\, f_X(x, v, t_0|x_0, v_0, t_0). \qquad (7.108)$$

Likewise, we can obtain the marginal pdf of the velocity $f_X(v, t|v_0, t_0)$ by integrating out the position degree of freedom.

Equation 7.107 is typically difficult to solve for a general potential $U(x)$. However, we can obtain a general solution in the stationary limit $t \to \infty$, where equilibrium is reached. In this limit, the lhs of Eq. 7.107 is zero, since $\lim_{t\to\infty} f_X(x, v, t|x_0, v_0) = f_X(x, v)$ and thus $\frac{\partial}{\partial t} f_X(x, v) = 0$. We then make a product ansatz:

$$f_X(x, v) \propto \phi(x) e^{-\frac{1}{2}\beta m v^2}. \tag{7.109}$$

Substituting this ansatz into Eq. 7.107 leads to

$$\begin{aligned} 0 = &-v e^{-\frac{1}{2}\beta m v^2} \phi'(x) - \beta v e^{-\frac{1}{2}\beta m v^2} U'(x) \phi(x) \\ &+ \frac{1}{m} \frac{\partial}{\partial v} \gamma v \phi(x) e^{-\frac{1}{2}\beta m v^2} + \frac{\Gamma}{2m^2} \frac{\partial}{\partial v} \beta m v \phi(x) e^{-\frac{1}{2}\beta m v^2}. \end{aligned} \tag{7.110}$$

Due to Eq. 7.18 the last two terms cancel and we obtain

$$\phi'(x) = -\beta U'(x) \phi(x). \tag{7.111}$$

The solution for $\phi(x)$ is thus

$$\phi(x) = e^{-\beta U(x)}, \tag{7.112}$$

and we obtain the stationary solution

$$f_X(x, v) = \mathcal{N} e^{-\frac{1}{2}\beta m v^2 - \beta U(x)}, \tag{7.113}$$

where $\mathcal{N}$ is a normalization constant that guarantees $\int_{-\infty}^{\infty} \mathrm{d}x \int_{-\infty}^{\infty} \mathrm{d}v\, f_X(x, v) = 1$.

We see that $f_X(x, v)$ is just the Boltzmann equilibrium distribution of a particle moving in the potential $U(x)$. This confirms that Eq. 7.1 (or Eqs. 7.101 and 7.102) correctly describe the fluctuations of a system in contact with a thermal reservoir. For general initial conditions x_0 and v_0 the system is initially out-of-equilibrium, but reaches its equilibrium state for $t \to \infty$.

7.4 The Overdamped Limit

The full time evolution of a system in its position-velocity phase space as described by Eq. 7.107 is usually difficult to deal with analytically. Here, we discuss an important parameter regime, where the velocity degrees of freedom can be neglected and a description of the system in terms of its position coordinate is sufficient. This is possible, when the velocity relaxes on a very fast time scale (adiabatically), such that only the evolution of the position is relevant. This limit is referred to as *overdamped limit,* since it physically represents a scenario of very low inertia or high friction, such that any inertial movement is instantaneously relaxed, similar to a particle moving in a highly viscous fluid.

In order to discuss this limit more quantitatively, we revisit force-free Brownian motion in position-velocity space. The equation of motion is

$$\dot{X}(t) = V(t) \tag{7.114}$$

$$m\dot{V}(t) = -\gamma V(t) + \xi(t). \tag{7.115}$$

As before, we take $\tau_r = m/\gamma$ as the characteristic relaxation time of the velocity. At equilibrium the initial velocity is given by $v_0^2 = \left\langle V^2 \right\rangle_{\text{eq}} = \frac{\Gamma\tau_r}{2m^2} = \frac{1}{m\beta}$, so that the velocity has the equilibrium correlation function (from Eq. 7.15)

$$\left\langle V(t)V(t') \right\rangle_{\text{eq}} = \frac{1}{m\beta} e^{-|t-t'|/\tau_r}. \tag{7.116}$$

The overdamped limit is characterized by $\tau_r \to 0$, which corresponds to a dynamical regime, where the velocity relaxes instantaneously either due to small inertia or large friction. In this limit

$$\lim_{\tau_r \to 0} \frac{1}{\tau_r} e^{-|t-t'|/\tau_r} = 2\delta(t-t'). \tag{7.117}$$

Therefore,

$$\left\langle V(t)V(t') \right\rangle_{\text{eq}} \to \frac{2}{\gamma\beta}\delta(t-t') \tag{7.118}$$

as $\tau_r \to 0$. Since $\left\langle V(t) \right\rangle = 0$ and $V(t)$ is a stationary process, we can replace $V(t)$ in Eq. 7.114 in the $\tau_r \to 0$ limit by a random force $\tilde{\xi}(t)$

$$\dot{X} = \tilde{\xi}(t). \tag{7.119}$$

As before, we use the notation $\tilde{\xi}(t)$ to indicate that this random force is actually a random velocity. It still has the properties Eqs. 7.8–7.10, but now with the noise strength $\tilde{\Gamma} = \frac{1}{\gamma\beta}$.

Therefore, the two dimensional problem Eqs. 7.114–7.115 reduces to a single degree of freedom for $\tau_r \to 0$: Eq. 7.119 is the overdamped equation of motion for the position coordinate. The reduction in the degrees of freedom by eliminating a fast degree of freedom (the velocity) is called adiabatic elimination. The random force is now on the level of the velocity instead of the force, which implies that velocities in the problem are no longer well defined.

Note that Eq. 7.119 describes a so called *Wiener process*, which represents a pure diffusion process of the position coordinate. The Fokker-Planck equation associated with the Wiener process is obtained from Eq. 7.93, which yields the diffusion equation

$$\frac{\partial}{\partial t} f_X(x, t|x_0, t_0) = \frac{\tilde{\Gamma}}{2}\frac{\partial^2}{\partial x^2} f_X(x, t|x_0, t_0) \tag{7.120}$$

for an initial condition $f_X(x, t_0|x_0, t_0) = \delta(x - x_0)$. The solution of the diffusion equation is a Gaussian centred at x_0 with variance $\tilde{\Gamma}(t - t_0)$:

$$f_X(x, t|x_0, t_0) = \frac{1}{\sqrt{2\pi\tilde{\Gamma}(t - t_0)}} e^{-\frac{1}{2}\frac{(x-x_0)^2}{\tilde{\Gamma}(t-t_0)}}. \tag{7.121}$$

The mean-square displacement is thus linear

$$\left\langle (X(t) - x_0)^2 \right\rangle = 2D(t - t_0). \tag{7.122}$$

We see that the diffusion coefficient is identified as $D = \frac{\tilde{\Gamma}}{2} = \frac{1}{\gamma\beta}$, which is just the Einstein relation Eq. 7.25. The overdamped equation of motion (7.119) thus correctly describes a diffusion process in the position coordinate.

We see that the result Eq. 7.119 can be obtained directly from Eq. 7.115 by setting $m = 0$, that is, neglecting the inertia. This procedure is possible since the noise acts additively in our framework. In the general case of a potential, we obtain the overdamped equation of motion

$$\dot{X}(t) = -\frac{U'(X)}{\gamma} + \tilde{\xi}(t), \tag{7.123}$$

by setting similarly $m = 0$ in Eq. 7.1 and $\tilde{\xi}(t)$ has the properties (7.8)–(7.10) with $\tilde{\Gamma} = \frac{2}{\gamma\beta}$. This equation is the starting point to describe diffusion processes, where inertia can be neglected and is therefore commonly applied to problems in highly viscous environments. The Fokker-Planck equation associated with Eq. 7.123 is then (from Eq. 7.93)

$$\begin{aligned}\frac{\partial}{\partial t} f_X(x, t|x_0, t_0) &= \frac{1}{\gamma}\frac{\partial}{\partial x}(U'(x) f_X(x, t|x_0, t_0)) \\ &\quad + \frac{\tilde{\Gamma}}{2}\frac{\partial^2}{\partial x^2} f_X(x, t|x_0, t_0),\end{aligned} \tag{7.124}$$

with initial condition $f_X(x, t|x_0, t_0) = \delta(x - x_0)$. Equation 7.124 is also called *Smoluchowski equation*. Note that Eq. 7.124 can be written as

$$\frac{\partial}{\partial t} f_X(x, t|x_0, t_0) = \frac{1}{\beta\gamma}\frac{\partial}{\partial x} e^{-\beta U(x)} \frac{\partial}{\partial x} e^{\beta U(x)} f_X(x, t|x_0, t_0). \tag{7.125}$$

Moreover, we can write Eq. 7.125 in the form

$$\frac{\partial}{\partial t} f_X(x, t|x_0, t_0) = -\frac{\partial}{\partial x} j(x, t|x_0, t_0) \tag{7.126}$$

$$j(x, t|x_0, t_0) = -\frac{1}{\beta\gamma} e^{-\beta U(x)} \frac{\partial}{\partial x} e^{\beta U(x)} f_X(x, t|x_0, t_0). \tag{7.127}$$

The quantity $j(x, t|x_0, t_0)$ is interpreted as probability current. Equation 7.126 highlights that the Fokker-Planck equation can be considered as a continuity equation for the probability.

We can verify that Eq. 7.124 leads to the correct equilibrium state by calculating the stationary solution $\lim_{t\to\infty} f_X(x, t|x_0, t_0) = f_X(x)$. In the stationarity limit, the lhs of Eq. 7.125 is zero so that

$$0 = \frac{1}{\beta\gamma}\frac{\partial}{\partial x} e^{-\beta U(x)} \frac{\partial}{\partial x} e^{\beta U(x)} f_X(x). \tag{7.128}$$

The integration constant is the stationary probability current $j(x, t|x_0, t_0) = \text{const}$, which is zero at equilibrium. Therefore

$$\frac{1}{\beta\gamma} e^{-\beta U(x)} \frac{\partial}{\partial x} e^{\beta U(x)} f_X(x) = 0, \tag{7.129}$$

which leads to

$$f_X(x) = \mathcal{N} e^{-\beta U(x)}. \tag{7.130}$$

As before, the normalization constant $\mathcal{N}$ has to be determined from the condition $\int_{-\infty}^{\infty} \mathrm{d}x\, f_X(x) = 1$. Equation 7.130 is again the correct Boltzmann equilibrium distribution of the microstate probabilities in the absence of the velocity degrees of freedom.

7.4.1 *Escape from a Metastable Potential*

Many interesting physical effects arise due to the effect of large fluctuations that move the system far away from its deterministic

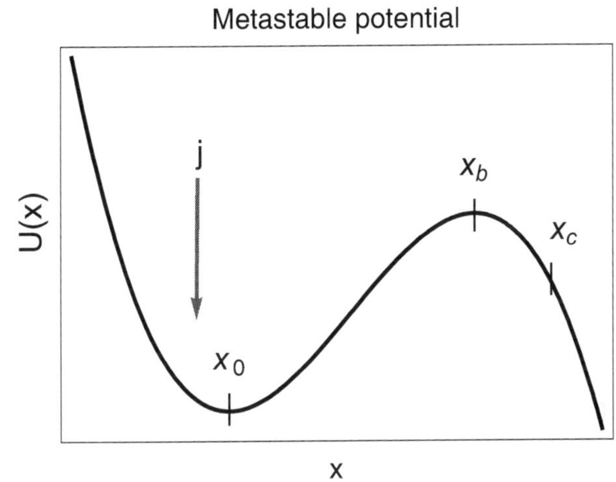

Figure 7.2 Plot of a metastable potential. The points x_0, x_b, and x_c used in the calculation are indicated. In the flux over population method a constant insertion of particles into the metastable basin at rate j is assumed.

attractor. An ubiquitous scenario is that of escape from a metastable state, which underlies a large variety of phenomena in physics, chemistry, and biology. Starting with the work by Kramers (Kramers, 1940), the theoretical study of the escape problem in thermal systems has attracted an extraordinary amount of attention over the years. For simplicity, we restrict the discussion here to a one-dimensional potential $U(x)$ with a single exit point of the metastable basin at x_c (see Fig. 7.2). We discuss the overdamped dynamics of a particle in this potential subject to thermal fluctuations. This motion is described by Eq. 7.123.

In the absence of any thermal noise, a particle inserted in the potential in the vicinity of x_0 will simply relax to x_0. The characteristic time of the relaxation is $\tau_r = \gamma / U''(x_0)$, which follows by approximating the potential around x_0 in a Taylor series up to quadratic orders. In the presence of the thermal fluctuations $\tilde{\xi}(t)$, the particle will be able to leave the metastable basin and escape via the exit point x_c. We are interested in a regime, where such an escape event is a rare event, i.e., the particle will spend a long time in the basin, before a rare sequence of fluctuations carries it across the barrier. This regime is realized either when the noise is weak (for low temperatures) or the barrier $\Delta U = U(x_b) - U(x_0)$ is large. Since the noise strength $\tilde{\Gamma} = \frac{2}{\gamma\beta}$, we can characterize the rare escape regime by $\beta\Delta U \gg 1$. In such a case the escape rate is well-defined and can be calculated using, e.g., the flux over population method first introduced by Kramers (Kramers, 1940).

Imagine a continuous insertion of particles by a constant current j into the metastable basin close to x_0. Over times $\gg \tau_r$, the system will settle into a stationary non-equilibrium state, where the particle insertion is balanced by the escape over the potential barrier and a stationary distribution develops inside the basin. The probability that an inserted particle is still in the metastable basin is

$$n_0 = \int_{-\infty}^{x_c} \mathrm{d}x\, f_X(x), \tag{7.131}$$

where $f_X(x)$ is the stationary pdf generated by Eq. 7.123 under such boundary conditions. The escape rate is then escape rate is then defined as

$$k = \frac{j}{n_0}. \tag{7.132}$$

In order to calculate k, we thus need to determine $f_X(x)$. The pdf is the stationary solution of the Smoluchowski Eq. 7.123 under the condition of a constant probability current $j(x, t|x_0, t_0) = j$ and the boundary condition $f_X(x_c) = 0$, i.e., we assume that every particle that moves across the potential barrier and reaches x_c indeed escapes the basin. In order to solve the Smoluchowski equation, we use again the representation Eq. 7.125. In the stationary limit the lhs of Eq. 7.125 is zero. Under the condition of a stationary current we have

$$\frac{1}{\beta\gamma} e^{-\beta U(x)} \frac{\partial}{\partial x} e^{\beta U(x)} f_X(x) = -j. \tag{7.133}$$

For the boundary condition $f_X(x_c) = 0$, we obtain then

$$f_X(x) = j\gamma\beta e^{-\beta U(x)} \int_x^{x_c} \mathrm{d}y\, e^{\beta U(x)}. \tag{7.134}$$

Therefore, we obtain the probability n_0 in the form of a double integral

$$\begin{aligned} n_0 &= \int_{-\infty}^{x_c} \mathrm{d}x\, f_X(x) \\ &= j\gamma\beta \int_{-\infty}^{x_c} \mathrm{d}x\, e^{-\beta U(x)} \int_x^{x_c} \mathrm{d}y\, e^{\beta U(y)}, \end{aligned} \tag{7.135}$$

and the escape rate Eq. 7.132 is

$$\frac{1}{k} = \gamma\beta \int_{-\infty}^{x_c} \mathrm{d}x\, e^{-\beta U(x)} \int_{x}^{x_c} \mathrm{d}y\, e^{\beta U(y)}. \tag{7.136}$$

The double integral can be evaluated by noting that the dominant contributions to the integrals come from the regions close to the extrema at x_0 and x_b. A Taylor expansion of the potential up to quadratic orders around x_c yields, e.g., for the second integral

$$\begin{aligned}\int_{x}^{x_c} \mathrm{d}y\, e^{\beta U(y)} &\approx \int_{-\infty}^{\infty} \mathrm{d}y\, e^{\beta U(x_b) - \frac{1}{2}\beta|U''(x_b)|(y-x_b)^2} \\ &= \frac{1}{\sqrt{2\pi\beta|U''(x_b)|}} e^{\beta U(x_b)},\end{aligned} \tag{7.137}$$

and similarly for the remaining integral, which is dominated by the region around x_0. Performing the integrations in this way we obtain

$$k = \frac{\sqrt{|U''(x_b)|U''(x_0)}}{2\pi\gamma} e^{-\beta\Delta U}, \tag{7.138}$$

which is Kramer's classical result for the escape rate in the overdamped regime. We see that the escape rate is small due to the dominant scaling $k \propto e^{-\beta\Delta U}$. This scaling of transition rates in thermally activated system has already been proposed by Van't Hoff and Arrhenius in the late 19th century.

7.5 Microrheology and the Generalised Langevin Equation

Our discussion so far has focused on the basic model Eq. 7.1 of a particle in a fluid environment subject to friction, a conservative force, and a thermal random force. This framework is general enough to capture the physics of microrheology with optical tweezers in Newtonian fluids. The effect of the optical tweezers is typically modelled in terms of a harmonic potential

$$U(x) = \frac{1}{2}\kappa x^2, \tag{7.139}$$

where the strength κ is also denoted as trap stiffness. The forces exerted on the colloidal particle by the fluid are captured in the linear friction force and in the random force, whereby the fluctuation-dissipation relation Eq. 7.18 needs to hold in order to describe the correct equilibrium statistics. The measurement of the fluid's viscosity then relies on the Stokes relation Eq. 7.2, which allows to determine η from the particle's position data as discussed in detail in Section 7.5.1 below.

However, the standard framework is not applicable to viscoelastic (non-Newtonian) fluids. In this case, the dynamics of an immersed colloidal particle is not appropriately described by Eq. 7.1, because the friction term is coupled in a more complicated way to the environment. An extension of Eq. 7.1 known as *generalised Langevin equation* is discussed in Section 7.5.2.

7.5.1 *Measuring Viscosity in Newtonian Fluids*

For a purely viscous fluid, the viscosity η is related to the friction coefficient by the Stokes relation Eq. 7.2. A common method to determine γ and thus η starts from a measurement of the position correlation function of the particle trapped by the optical tweezers. In the presence of the potential Eq. 7.139 the dynamics of the colloidal particle is given by Eq. 7.26, where $\omega_0^2 = \kappa/m$. The position correlation function is thus determined in a straightforward way from the result Eq. 7.40 by calculating the matrix exponential or from Eq. 7.62 by calculating the inverse Fourier transform of the spectral density. The result in the stationary regime is

$$\langle X(\tau)X(0)\rangle_{\mathrm{eq}} = \frac{\lambda_2 e^{\lambda_1 \tau} - \lambda_1 e^{\lambda_2 \tau}}{\beta\kappa(\lambda_2 - \lambda_1)}, \tag{7.140}$$

where $\lambda_{1,2}$ are the eigenvalues of the matrix $\mathbb{M}$

$$\lambda_{1,2} = -\frac{1 \pm \sqrt{1 - 4\tau_r^2\omega_0^2}}{2\tau_r}. \tag{7.141}$$

In the overdamped limit $\tau_r \to 0$ we obtain then

$$\langle X(\tau)X(0)\rangle_{\mathrm{eq}} = \langle X^2\rangle_{\mathrm{eq}}\, e^{-\tau/\tau_c}, \tag{7.142}$$

with the time scale $\tau_c = \omega_0^2\tau_r = \kappa/\gamma$ governing the decay of the position correlations. Note that Eq. 7.142 can also be derived by integrating the overdamped equation of motion (7.123) analogous to the calculation of the velocity correlation function of the Ornstein-Uhlenbeck process in Section 7.2.1. With Eq. 7.2 we obtain the viscosity

$$\eta = \frac{\gamma}{6\pi a} = \frac{\tau_c \kappa}{6\pi a}, \tag{7.143}$$

so that from a knowledge of κ and τ_c the viscosity η can be determined. Due to the equipartition theorem, the trap stiffness is related to the variance of the particle position via

$$\kappa = \frac{1}{\beta \langle X^2 \rangle_{\text{eq}}}, \tag{7.144}$$

so that both κ and τ_c are determined from the position data.

7.5.2 *Viscoelasticity*

In order to take into account the viscoelasticity of the medium a generalised Langevin equation has been proposed for the dynamics of a colloidal particle in a non-Newtonian fluid (Mason and Weitz, 1995)

$$m\dot{V}(t) = -\int_0^t \mathrm{d}t' \zeta(t-t')V(t') - \kappa X(t) + \xi(t). \tag{7.145}$$

Here, the harmonic potential Eq. 7.139 is again used to model the optical trap. The main difference with the Newtonian case is the presence of the memory friction $\int_0^t \mathrm{d}t' \zeta(t-t')V(t')$, which captures the effect of energy stored in the medium during the time evolution. The thermal force $\xi(t)$ satisfies $\langle \xi(t) \rangle = 0$ as before, but differs in the covariance, which is now

$$\langle \xi(t)\xi(t') \rangle = \frac{1}{\beta}\zeta(t-t'). \tag{7.146}$$

Equation 7.146 is required in order for Eq. 7.145 to reach the correct equilibrium state, i.e., the Boltzmann distribution Eq. 7.113 in the stationary limit. Eq. 7.146 is thus also called the 2nd fluctuation dissipation theorem. In the absence of any memory effects we have $\zeta(t) = 2\gamma\delta(t)$, so that we recover the usual Langevin equation and noise covariance.

In the following, we discuss how the generalised Langevin equation can be used to determine the viscosity according to the original work of Ref. (Mason and Weitz, 1995). It is easier to work with Eq. 7.145 in Laplace space, where the convolution integral becomes a product. Introducing Laplace transforms $\hat{V}(s) = \int_0^\infty \mathrm{d}t\, e^{-ts} V(t)$, and likewise for the other time-dependent variables, we obtain

$$m(s\hat{V}(s) - V(0)) = \hat{\xi}(s) - \hat{\zeta}(s)\hat{V}(s) - \kappa\hat{X}(s). \tag{7.147}$$

Since $\dot{X}(t) = V(t)$, we can write $\hat{X}(s) = \hat{V}(s)/s$ (for $X(0) = x_0 = 0$). Solving Eq. 7.147 for $\hat{V}(s)$ yields then

$$\hat{V}(s) = \frac{mV(0) + \hat{\xi}(s)}{ms + \hat{\zeta}(s) + \kappa/s} \tag{7.148}$$

We then multiply Eq. 7.148 by $V(0)$ assuming that the initial velocity is at equilibrium, i.e., distributed according to the equilibrium distribution. Taking the equilibrium ensemble average yields

$$\left\langle \hat{V}(s)V(0)\right\rangle_{\text{eq}} = \frac{m\left\langle V^2\right\rangle_{\text{eq}} + \left\langle V(0)\hat{\xi}(s)\right\rangle_{\text{eq}}}{ms + \hat{\zeta}(s) + \kappa/s}. \tag{7.149}$$

Here, we can further use the property $\left\langle V(0)\hat{\xi}(s)\right\rangle_{\text{eq}} = 0$, since the random force is uncorrelated with realizations at previous times and the result $\left\langle V^2\right\rangle_{\text{eq}} = 1/(\beta m)$ from the equipartition theorem. This yields

$$\left\langle \hat{V}(s)V(0)\right\rangle_{\text{eq}} = \frac{1}{\beta(ms + \hat{\zeta}(s) + \kappa/s)}. \tag{7.150}$$

On the other hand, we can connect the velocity correlation function with the mean-square displacement in d dimensions using the relationship

$$\frac{\mathrm{d}^2}{\mathrm{d}\tau^2}\left\langle \mathbf{R}^2(\tau)\right\rangle_{\text{eq}} = 2d\, C_{vv}(\tau). \tag{7.151}$$

The derivation of Eq. 7.151 is discussed in the side remark below. With the Laplace transform of Eq. 7.151 we obtain further

$$\left\langle \hat{V}(s)V(0)\right\rangle_{\text{eq}} = \hat{C}_{vv}(s) = \frac{s^2\left\langle \widehat{\mathbf{R}^2}(s)\right\rangle}{2d}. \tag{7.152}$$

The solution for $\hat{\xi}(s)$ is then

$$\hat{\xi}(s) = \frac{2d}{\beta s^2 \left\langle \widehat{\mathbf{R}^2}(s) \right\rangle} - ms - \frac{\kappa}{s}. \tag{7.153}$$

We can finally relate the memory friction to the frequency dependent viscosity by an extension of the usual Stokes relationship Eq. 7.2

$$\hat{\xi}(s) = 6\pi a \,\hat{\eta}(s). \tag{7.154}$$

The final result is thus

$$\hat{\eta}(s) = \frac{1}{6\pi a} \left(\frac{2d}{\beta s^2 \left\langle \widehat{\mathbf{R}^2}(s) \right\rangle} - ms - \frac{\kappa}{s} \right). \tag{7.155}$$

As before the trap stiffness κ is determined from the equipartition theorem using Eq. 7.144. Equation 7.155 highlights that the central quantity in the non-Newtonian case is the d-dimensional MSD $\left\langle \mathbf{R}^2(\tau) \right\rangle_{\text{eq}}$, which exhibits typically a non-trivial time dependence. For comparison, the MSD in Newtonian fluids is always linear in time: $\left\langle \mathbf{R}^2(\tau) \right\rangle = 2dD\tau$.

Side remark: The equilibrium velocity correlation function is related to the MSD as follows. A straightforward extension of Eq. 7.21 to d dimensions yields

$$\left\langle \mathbf{R}^2(\tau) \right\rangle = \int_0^\tau \mathrm{d}t_1 \int_0^\tau \mathrm{d}t_2 \left\langle \mathbf{V}(t_1)\mathbf{V}(t_2) \right\rangle, \tag{7.156}$$

for an initial position $\mathbf{r}_0 = \mathbf{0}$. If the system is sufficiently isotropic we can express the d-dimensional velocity correlation function in terms of the velocity magnitude

$$\begin{aligned} \left\langle \mathbf{V}(t_1)\mathbf{V}(t_2) \right\rangle &= \left\langle V_x(t_1)V_x(t_2) \right\rangle + \left\langle V_y(t_1)V_y(t_2) \right\rangle + \left\langle V_z(t_1)V_z(t_2) \right\rangle \\ &= 3 \left\langle V(t_1)V(t_2) \right\rangle, \end{aligned} \tag{7.157}$$

e.g., in $d = 3$ dimensions. In the stationary regime we can thus write

$$\begin{aligned} \left\langle \mathbf{R}^2(\tau) \right\rangle_{\text{eq}} &= d \int_0^\tau \mathrm{d}t_1 \int_0^\tau \mathrm{d}t_2 \left\langle V(t_1)V(t_2) \right\rangle_{\text{eq}} \\ &= d \int_0^\tau \mathrm{d}t_1 \int_0^\tau \mathrm{d}t_2 C_{vv}(t_2 - t_1) \\ &= 2d \int_0^\tau \mathrm{d}t_2 \int_0^{t_2} \mathrm{d}t_1 C_{vv}(t_2 - t_1). \end{aligned} \tag{7.158}$$

The last step follows since $C_{vv}(t_2 - t_1)$ is a function of $|t_2 - t_1|$. Therefore, we obtain Eq. 7.151 by taking derivatives

$$\frac{\mathrm{d}^2}{\mathrm{d}\tau^2}\left\langle \mathbf{R}^2(\tau)\right\rangle_{\mathrm{eq}} = 2d\, C_{vv}(\tau).$$

Chapter 8

Most Popular Microrheology Techniques

Aristeidis Papagiannopoulos
Theoretical and Physical Chemistry Institute, National Hellenic Research Foundation, 48 Vass. Constantinou Ave., 11635 Athens, Greece
apapagiannopoulos@eie.gr

8.1 Introduction

During the last twenty years microrheology methods have emerged as sensitive tools for the investigation of materials' mechanical properties at micro- and nano-length scales. Although the materials' bulk properties are within their main targets, microrheology techniques probe the materials' dynamics at a microscopic level, providing potentially a more direct link between the material's (micro- and nano-) structure and its mechanical properties. In general, microrheology techniques are aimed at uncovering the linear viscoelastic properties of materials by means of statistical mechanical analysis of the trajectory of micron sized probe particles suspended in the material under investigation.

The thermally driven *random* motion of micron-sized particles in fluids was first observed by Robert Brown almost 200 years ago. Then the correct theoretical explanation and mathematical

Microrheology with Optical Tweezers: Principles and Applications
Edited by Manlio Tassieri

ISBN 978-981-4669-18-4 (Hardcover), 978-981-4669-19-1 (eBook)
www.panstanford.com

formulation came with the formulation of the famous *fluctuation-dissipation theorem* and Einstein's work on statistical mechanics. The random collisions of fluids molecules on the particle cause the random displacement of the particle that is dissipated by the collisions with the fluid's molecules themselves. Therefore, the thermally driven random motion of the probe particle inside the fluid was connected to the mechanical properties of the fluid. Indeed, the pioneering article of Mason and Weitz (Mason and Weitz, 1995) gave birth to the field of Microrheology. In their work, Mason and Weitz introduced the generalised Stokes-Einstein relationship as a mathematical tool to connect the measured mean squared displacement of the probe particle with the viscoelastic properties of the fluid. The major advantages that microrheology techniques have offered so far, are the need of very small sample volumes (few microliters) and the extension of rheological measurements to very high frequencies (up to MHz), *plus* the opportunity of exploring single molecular dynamics.

In this chapter the most popular microrheology techniques are reviewed. The discussion is initially focused on the so called *passive* microrheology techniques, which are characterised by the fact that the motion of the suspended probe particles is *thermally* driven. Whereas, *active* microrheology techniques, for which the motion of the probe particles is driven by an externally applied force field, are introduced at the end of this chapter for the case of 'Magnetic Tweezers' and, in more detail, in Chapter 9 for the case of 'Optical Tweezers'. They can also be seen as a possible extension of passive microrheology techniques to highly viscous fluids or as an opportunity for exploring non-linear microrheology measurements.

8.2 Theoretical Background of Microrheology

A microscopic particle inside a fluid is *entrained* by the thermal (Brownian) fluctuations of the fluid's molecules that force it to move with a *random* trajectory. This motion can be described by the mean-squared-displacement (MSD, $\langle \Delta r^2 \rangle$) of the particle as a function of time (t) (or time interval $\tau = t - t_0$, where t_0 is the instant at which the observation starts). In the case of a Newtonian fluid (i.e. a fluid with constant viscosity η), these two quantities are simply proportional to each other:

$$\langle \Delta r^2(\tau) \rangle = 6D\tau \tag{8.1}$$

where D is the diffusion coefficient of the particle, which is related to both the particle's hydrodynamic radius (a) and the fluid's viscosity by means of the renowned Stokes-Einstein equation:

$$D = \frac{k_B T}{6\pi a \eta} \tag{8.2}$$

where k_B is the Boltzmann constant and T is the absolute temperature. For a compact and rigid spherical particle the hydrodynamic radius is identical to the particle's geometric radius. Therefore, a measurement of the diffusion coefficient of a spherical particle that diffuses inside a Newtonian fluid is equivalent to a measurement of the fluid's viscosity. In other words, the diffusing colloidal particle acts as a 'probe' for the measurement of the fluid's viscosity. By combining Eqs. 8.1 and 8.2 one obtains:

$$\langle \Delta r^2(\tau) \rangle = \frac{k_B T}{\pi a \eta} \tau \tag{8.3}$$

which explicits the relationship between the measured MSD and the viscosity of the suspending Newtonian fluid. While the rheological properties of a Newtonian fluid are identified by a single parameter, i.e. the viscosity, those of a generic complex fluid (i.e. a non-Newtonian fluid) can be represented by two components describing the *elastic* and the *viscous* nature of the fluid. These components are the *storage* and the *loss* moduli of the complex fluid, i.e., $G'(\omega)$ and $G''(\omega)$, respectively; as already introduced in Chapter 6.

The general concept of microrheology can be seen as an extension to non-Newtonian fluids of the viscosity measurement by means of the probe particle diffusion coefficient. Indeed, if the probe's random motion in a Newtonian fluid contains the rheological signature of the fluid's viscosity, then it is expected that the probe's random motion into a viscoelastic fluid should contain the rheological signature of both the viscoelastic moduli. A typical example of viscoelastic fluid is a solution of entangled polymer chains, as schematically shown in Fig. 8.1. In this case the probe particle *feels* the environment made of both the overlapping

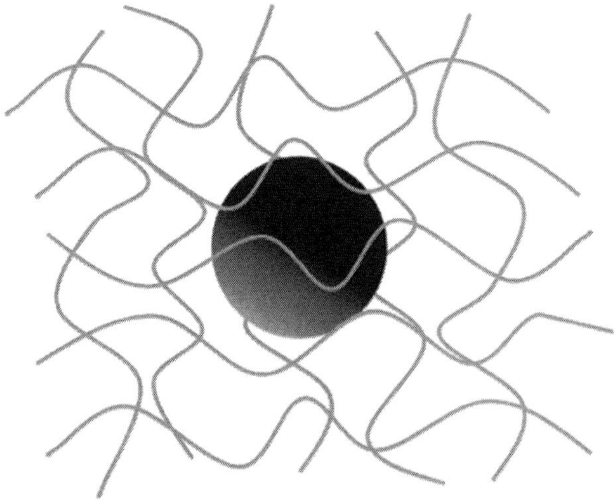

Figure 8.1 Schematic representation of a spherical colloidal particle immersed into an entangled polymer solution.

macromolecular chains and the solvent molecules as a whole (i.e. as a continuous medium).

The thermal motion of a colloidal particle inside a viscoelastic fluid can be *qualitatively* described in a straightforward way via two extreme cases: (i) a purely viscous fluid and (ii) a purely elastic solid. In a purely viscous fluid the particle will undergo diffusive motion as described by Eq. 8.3. In a purely elastic solid the particle will undergo confined random fluctuations, but it will not diffuse with time. The amplitude of the fluctuations will depend on the relative magnitude of the thermal energy and the elastic modulus of the solid. In particular, the MSD of a constrained probe particle into a purely elastic solid is directly proportional to the thermal energy and inversely proportional to material's elastic modulus (G'), with no dependency on time:

$$\langle \Delta r^2(\tau) \rangle \propto \frac{k_B T}{G'} \tag{8.4}$$

Graphically, the above two cases can be discriminated by analysing the power law of the MSD as function of time ($MSD \propto \tau^{\alpha}$) in a double logarithmic plot (i.e., $\log(MSD) \propto \alpha \log(\tau)$). In fact, for a purely viscous fluid this slope (α) must be equal to one, whereas for a purely elastic solid this slope must be equal to zero (at least at long times). It follows that, for a viscoelastic fluid the slope of the particle's MSD is expected to vary within the interval $\alpha \in [0, 1]$, so that, when $\alpha \cong 1$ a predominantly viscous behaviour is observed, whereas when $\alpha \approx 0$ a predominantly elastic behaviour is shown. The change of the slope as a function of time is a signature of the frequency-dependence of the material's rheological properties.

In order to derive a relationship between the output of a microrheology experiment (most commonly the MSD) and the desired rheological quantity (e.g. the shear complex modulus), Mason and Weitz (Mason and Weitz, 1995) employed a generalised Langevin equation:

$$m\dot{v}(t) = f_{\mathrm{R}}(t) - \int_0^t \zeta(t-\tau)v(\tau)d\tau \tag{8.5}$$

where m is the mass of the particle, $v(t)$ its velocity and $f_{\mathrm{R}}(t)$ is the resultant of all random forces acting on the particle. The viscous damping of the fluid is represented by an integral term incorporating the microscopic memory function $\zeta(t)$. In order to solve Eq. 8.5, some considerations and assumptions were necessary. In particular, the fluctuation-dissipation theorem for the random force acting on a particle suspended into a viscoelastic fluid gives:

$$\langle f_{\mathrm{R}}(0) \cdot f_{\mathrm{R}}(t) \rangle = k_{\mathrm{B}} T \zeta(t) \tag{8.6}$$

Moreover, the particle's velocity autocorrelation function can be expressed in terms of its MSD and by assuming that the Laplace-transformed bulk viscosity of the fluid $\tilde{\eta}(s)$ is proportional to the microscopic memory function $\tilde{\zeta}(s) = 6\pi a \tilde{\eta}(s)$, one can obtain the expression of the complex shear modulus in the Laplace domain:

$$\tilde{G}(s) = s\tilde{\eta}(s) = \frac{s}{6\pi a}\left(\frac{6k_{\mathrm{B}}T}{s^2 \left\langle \widetilde{\Delta r^2}(s) \right\rangle} - ms\right) \tag{8.7}$$

where $\left\langle \widetilde{\Delta r^2}(s) \right\rangle$ is the Laplace transform of the particle's MSD and s is the Laplace frequency. Notice that, the second term in the right hand side of Eq. 8.7 is the inertia term that for micron-sized particles becomes significant only at relatively high frequencies (i.e. of the order of MHz) and it is commonly neglected in microrheology measurements. This leads to the Generalised Stokes-Einstein Relationship (GSER) in the Laplace domain, which connects the particle's MSD to the shear modulus of a generic viscoelastic fluid:

$$\tilde{G}(s) = \frac{k_B T}{\pi a s \left\langle \widetilde{\Delta r^2}(s) \right\rangle} \tag{8.8}$$

In order to determine $\tilde{G}(s)$ the Laplace transform of the experimentally measured MSD is thus needed. A common method for determining the Laplace transform of the measured MSD is based on a numerical approximation that works on the qualitative information embedded into the local slope of the log-log representation of the MSD (Mason, 2000). In particular, the bypass to the problematic Laplace transform of a finite set of experimental data points in a limited time window is achieved by considering a local transformation of the data. In a log-log plot, the local behaviour of the MSD, around a time point $t = 1/s$, can be assumed linear, i.e.:

$$\log\left\langle \Delta r^2(\tau) \right\rangle \approx \log\left\langle \Delta r^2(1/s) \right\rangle + a(1/s) \cdot (\log(\tau) - \log(1/s)) \tag{8.9}$$

The slope $a(1/s)$ is numerically calculated, normally by finite differences or by fitting sufficiently smooth functions around the point of interest, making sure that the general trend of the curve is followed over a sufficiently large range of data. The resulting local power-law,

$$\left\langle \Delta r^2(\tau) \right\rangle \approx \left\langle \Delta r^2(1/s) \right\rangle \cdot (st)^{a(1/s)} \tag{8.10}$$

has a known Laplace transform:

$$s \cdot \left\langle \widetilde{\Delta r^2}(s) \right\rangle \approx \left\langle \Delta r^2(1/s) \right\rangle \cdot \Gamma\left(1 + a(1/s)\right) \tag{8.11}$$

where Γ denotes the Gamma function. This expression assumes that contributions to the transformation integral for the behaviour of MSD at times much different (logarithmically) than $1/s$ can be effectively neglected. This approximation works very well when the MSD is sufficiently smooth for the power law expansion not to be a rough simplification.

The final passage to the frequency domain is simply obtained by swapping the Laplace frequency s with the Fourier frequency ω and the complex shear modulus and its components can be expressed as:

$$|G^*(\omega)| = \frac{k_B T}{\pi a \langle \Delta r^2(1/\omega) \rangle \cdot \Gamma(1 + a(1/\omega))} \tag{8.12}$$

$$G'(\omega) = |G^*(\omega)| \cos\left(\frac{\pi a(1/\omega)}{2}\right) \tag{8.13}$$

$$G''(\omega) = |G^*(\omega)| \sin\left(\frac{\pi a(1/\omega)}{2}\right). \tag{8.14}$$

The resulting expressions retain the physical content of the fluid's viscoelastic behaviour. Indeed, if the slope $a(t)$ is greater than $1/2$ the viscous behaviour dominates; whereas, if it is lower than $1/2$ the elastic behaviour dominates.

An alternative representation of the fluid's viscoelastic properties by means of the particle's MSD is obtained by considering the relationship between the fluid's creep compliance and the particle's MSD (Xu et al., 1998):

$$J(t) = \frac{\pi a}{k_B T} \langle \Delta r^2(\tau) \rangle. \tag{8.15}$$

This expression is very useful because the creep compliance is a bulk-rheology function directly measured in the time-domain and no Fourier or Laplace transforms of the MSD are needed. However, the relationship between the time-dependent shear creep compliance and the frequency-dependent complex shear modulus have been described in Chapter 6.

The analytical treatment of the particle's MSD introduced so far in based on the assumption that the particle motion is thermally driven (i.e. free diffusive motion). This can also be seen as an induced motion by means of the *weakest* driving force that can be achieved in microrheology. Moreover, this has the great advantage that both the stresses and the strains involved in passive microrheology measurements have extremely small magnitudes and therefore they fall into the linear regime of the material under investigation. In the past few years many passive microrheology techniques have been developed and the most popular of these are presented hereafter. In the last paragraph of this chapter, an example of active microrheology with magnetic tweezers is described.

8.3 Video Particle Tracking Microrheology

A straightforward method for measuring the MSD of colloidal particles suspended into a fluid is via Video Particle Tracking (VPT). The main advantages of this technique are (i) the use of a small amount of sample (i.e. tens of μl or less, depending on the sample holder) and (ii) the unique capability of tracking individual particles instead of measuring averages of particle ensembles. A typical experimental set-up consists of an optical microscope (or in cases of opaque samples a fluorescent microscope combined with fluorescent probe particles) connected to a camera (usually equipped with a charge-coupled device (CCD)) that allows visualisation of the colloidal spheres' motion inside the specimen, as schematically shown in Fig. 8.2. Monodisperse in size spherical colloidal particles of known radius (usually ranging from 250 nm to 500 nm) are immersed into the fluid at a concentration such that an average number of 50 to 100 particles are in focus at the same time during the measurement. The images can be stored either on a digital video disc (by means of a DVD recorder) or on a personal computer via a frame grabber. The frame acquisition rate defines the highest frequency accessible with VPT experiments. Whereas, the lowest frequency is given by the inverse of the mean permanence time of the probe particles within the field of view; usually this time is of the order of few seconds. In the generic set-up of Fig. 8.2 the sample, $20-30$ μl in volume, is loaded on a slide with a cavity having a depth of circa 1 mm and a diameter of circa 10 mm. The filled cavity is then covered with a glass coverslip of thickness $\approx$100 μm. A 100X oil immersion lens with $\approx$200 μm working distance is used to focus into the sample well. Length scales in VPT are readily calibrated by means of a stage micrometer and the resulting magnification on the camera is typically of the order of 0.2–0.4 μm/pixel. Movies of diffusing particles are typically taken for a period of time of few minutes, with a compromise on the duration based on the balanced between a statistically valid measurement and the amount of storage space available for the movie. The concentration of the particles is optimised so that the resulting images contain enough probes for a good statistics, but not too many to affect either the particle tracking procedure or to induce undesired particle interactions. An

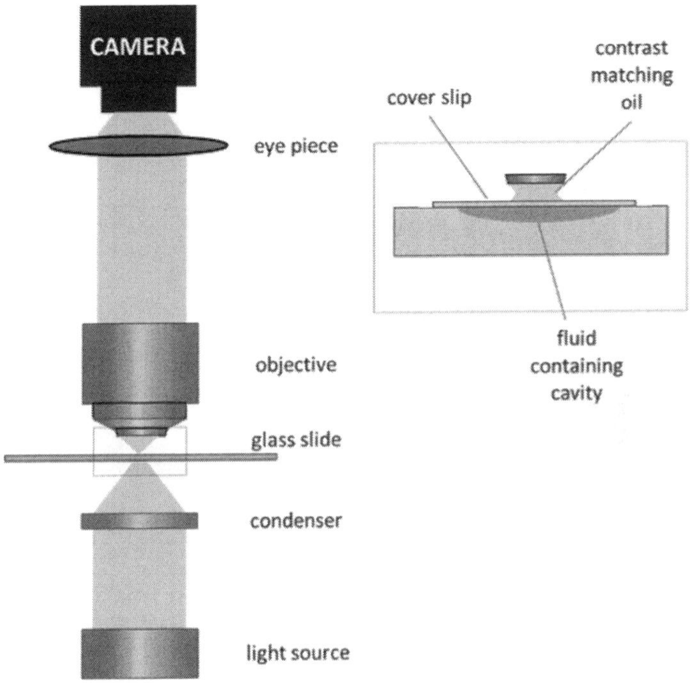

Figure 8.2 Schematic representation of an experimental set-up for Video Particle Tracking microrhelogy.

image analysis software (e.g. IDL or MATLAB) is used to extract the trajectories of the particles from the recorded movies. The movies consist of a series of successive frames. Therefore, in order to draw the trajectories of all the particles observed during the movie, a particle tracking algorithm has to locate all the particles in each image and then connect the positions of every particle between successive images. The basic principles underpinning the image analysis process for particle tracking of colloidal systems have been masterly introduced by Crocker and Grier (Crocker and Grier, 1996) and they are following summarised. The 'raw-data' of a VPT experiment is a movie of diffusing particles. The series of frames that make up the movie are readily obtained by a video analysis software except from the case of a frame grabber where the frames are sequentially obtained from the camera and stored onto the hard disk. Each frame is a 2-D array that contains the brightness $A(x, y)$ of the corresponding pixels on the x, y plane of view. The brightness array is convoluted by the kernel $K(i, j)$ in order to remove noise due to contrast gradients in the image or modulation of the background brightness and digitisation noise from the frame grabber or CCD camera:

$$A_{\text{conv}}(x, y) = \sum_{i,j=-w}^{w} K(i, j) \cdot A(x + i, y + j) \tag{8.16}$$

where w is an integer with a value between the single particle's apparent radius and the average inter-particle separation distance and

$$K(i, j) = \frac{1}{B} \exp\left(-\frac{(i^2 + j^2)}{4\lambda_n^2}\right) - \frac{1}{(2w + 1)^2}. \tag{8.17}$$

In Eq. 8.17 the Gaussian term corrects for the digitisation noise and the correlation length λ_n is normally taken equal to 1 pixel. The other term performs the subtraction of the long-wavelength noise background, averaging over squares of side $2w$. The normalisation factor is:

$$B = \left(\sum_{-w}^{w} \exp\left(-\frac{i^2}{4\lambda_n^2}\right)\right)^2. \tag{8.18}$$

After the background correction, local maxima in brightness are chosen as candidate particles. A point is chosen as a particle center if it has the maximum brightness within distance w. The particles' position can be estimated with an uncertainty even lower than $1/10$ of a pixel. This is due to the refinement of the particles position using the centroid calculation of Eq. 8.19. If the initial position is (x, y) the centroid offset (ϵ_x, ϵ_y) is calculated and the refined center is given by $(x + \epsilon_x, y + \epsilon_y)$. The procedure is repeated until the offsets' absolute values do not exceed 0.5 pixels.

$$(\epsilon_x, \epsilon_y) = \frac{1}{m} \sum_{(i^2+j^2)\leqslant w^2} (x, y) \cdot A(x + i, y + j) \tag{8.19}$$

where

$$m = \sum_{(i^2+j^2)\leqslant w^2} A(x + i, y + j). \tag{8.20}$$

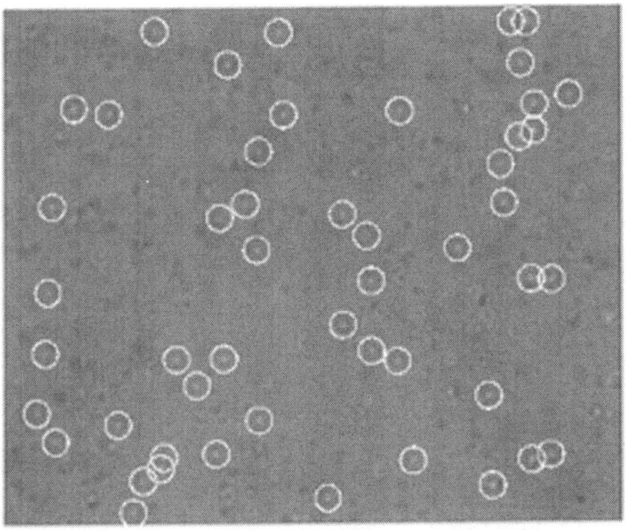

Figure 8.3 Image (352 × 288 pixels) of polystyrene spheres suspended in a guar solution at concentration of 0.2% w/w. The features selected as 'good' particles by the image analysis software are indicated with a white circle around them. Image reproduced from (Papagiannopoulos, 2005).

The final step in particle tracking is to connect the particles' positions of every frame to the positions in the next frame in order to create the individual trajectories. All the possible bonds between the particles of a frame with the particles of the next frame are formed. In order to make the calculations more efficient, only the bonds shorter than a length L are taken into account. This length is chosen to be lower than half of the average inter-particle distance and higher than the typical single particle displacement between successive frames. The bonds that identify particles' motion from one frame to the other are the ones that minimise the sum $\sum_{i=1,M} \delta_i^2$ within each subset of bonds of number M with bond lengths δ_i.

A snapshot of 0.5 μm radius polystyrene spheres in water is shown in Fig. 8.3. The bright particles chosen by the image analysis software are shown with a white circle around them. In Fig. 8.4, the trajectories of a single polystyrene sphere suspended in both water and water-based solution of guar are shown. The trajectory in the guar solution is clearly less expanded than in water since the guar

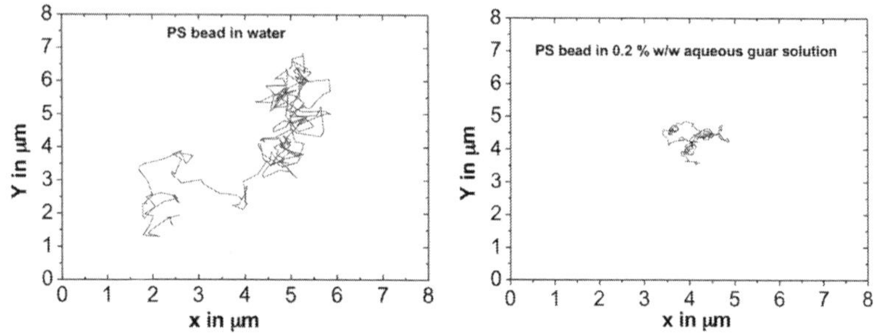

Figure 8.4 The trajectories over a period of ten seconds of PS particles suspended in water (left) and water-based solution of guar (right). Graphs reproduced from (Papagiannopoulos, 2005).

solution is a fluid with significantly higher viscosity than water. Of course, the observed motion in VPT is a 2-D projection, on the focus plane of the objective lens, of a 3-D trajectory. However, the MSD of an isotropic 3-D random walk projected onto a 2-D plane is simply scaled by a constant factor equal to $2/3$.

The output of the tracking process is an array containing the (x, y) positions of every particle in every frame. The frame acquisition rate of the camera defines the time label of each frame. A camera that performs at 25 Hz provides 25 frames in one second and therefore the time between successive frames is 0.04 seconds. The mean squared displacement is calculated by:

$$\langle \Delta r^2(\tau) \rangle = \langle [\vec{r}(t+\tau) - \vec{r}(t)]^2 \rangle_{t,N} \tag{8.21}$$

where t is the absolute time, $\vec{r}(t)$ the position of an individual particle and the average is taken over all initial times t and particles present in the experiment. Time τ is the lag-time, i.e. the time difference from the time t taken as initial. Under the ergodicity assumption one can average between different ensembles for the same initial real time or equivalently on the same ensemble over different initial real times.

In Fig. 8.5 (left) the $\langle \Delta r^2(\tau) \rangle$ of polystyrene spheres in an aqueous solution of polystyrene sulfonate comb polyelectrolytes at several polyelectrolyte concentrations are presented. The displacements of the particles decrease as the concentration increases due to the increase in viscosity of the solutions. Notably, VPT shows

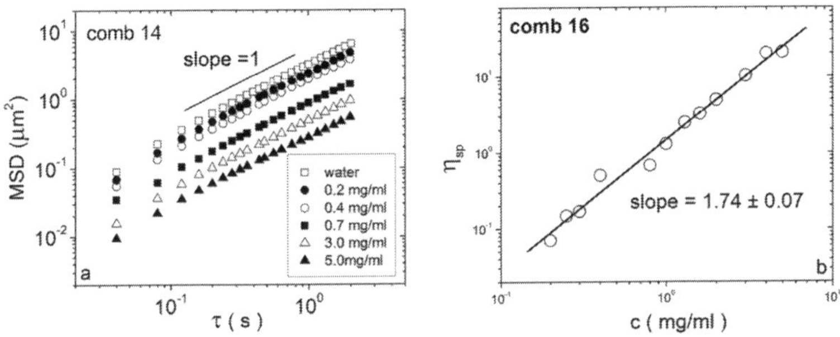

Figure 8.5 (Left) MSDs of 0.548 μm polystyrene particles as a function of time for sparse polyelectrolyte comb in water with no added salt. Several polymer concentrations are shown, including the case of pure solvent. A line with slope 1 indicates the diffusive behaviour of the spheres in the solutions. (Right) Specific viscosity of a dense polyelectrolyte comb as a function of polymer concentration in aqueous solutions with no added salt measured by VPT (straight line is a power-law fit). Graphs reproduced from (Papagiannopoulos et al., 2005a).

a great sensitivity in measuring small viscosity changes vey near to that of the solvent (i.e. at very low solute concentration). The dependence of the specific viscosity $\eta_{\text{sp}} = (\eta_{\text{solution}} - \eta_{\text{solvent}})/\eta_{\text{solvent}}$ on concentration is shown in Fig. 8.5 (right); from which it is possible to evaluate the concentration scaling law of η_{sp}. This can be then related to the molecular structure and interactions by means of theoretical models.

One of the aims of microrheology is the investigation of local viscoelasticity in living cells and inhomogeneous biological fluids (Wirtz, 2009). This unique capability is naturally provided by VPT and it has been presented in the pioneering works of Denis Witrz (Wirtz, 2009). Moreover, VPT has been used for the statistical analysis and mapping of mechanically inhomogeneous environments (Apgar et al., 2000) and the principal role of actin filaments in cell mechanics have been quantified in situ and in vitro.

As mentioned in the introduction the GSER can be used to connect the MSD of particles with the viscoelastic response of the fluid. However, this equation is subject to a number of assumptions including non-slip boundary conditions and incompressibility. There is also a limited range of frequencies over which the experiments can be applied. The lower limit is set by the fact that the motion of solvent and polymer matrix are no longer coupled (Levine and Lubensky, 2000). Whereas, the higher limit is related to the onset of the effects of the particle's inertia.

In video particle tracking there are two kinds of uncertainty that affect the measurements, i.e. the static error and the dynamic error. The particle displacements become smaller as the viscoelastic moduli of the samples increase and eventually are affected by the resolution (ϵ) of the experimental setup and tracking algorithms. The effect of the static error is stronger at short time scales (or high frequencies) where normally the MSDs are small. The dynamic error is caused by the finite shutter time (σ) of the specific camera and results to an under-estimation of the actual displacement (Savin and Doyle, 2005). In Fourier space the effect of these two errors can be expressed as:

$$\left\langle \widehat{\Delta r^2}(\omega, \sigma) \right\rangle_{\text{exp}} = \frac{\sin^2(\omega\sigma/2)}{(\omega\sigma/2)^2} \left\langle \widehat{\Delta r^2}(\omega, \sigma) \right\rangle + 2\epsilon^2 \tag{8.22}$$

where $\left\langle \widehat{\Delta r^2}(\omega, \sigma) \right\rangle_{\text{exp}}$ and $\left\langle \widehat{\Delta r^2}(\omega, \sigma) \right\rangle$ are the Fourier transforms of the measured MSD $\langle \Delta r^2(\tau) \rangle_{\text{exp}}$ and of the actual particle MSD $\langle \Delta r^2(\omega, \sigma) \rangle$, respectively. The first term on the right side of Eq. 8.22 contains the dynamic error, which depends on both the frequency ω and the exposure or shutter time σ. Normally it affects only the first few data points of the measurement and it can be neglected. The other term is the static error and is simply represented by a constant baseline.

While microrheology thechniques are meant to offer an alternative to conventional bulk-rheology methods, their microscopic nature may affect the results. This because, over the duration of an experiment, a microscopic particle can experience only a finite volume (i.e. a small portion of the local environment surrounding it) of the suspending fluid. Therefore, microrheology measurements reflect the bulk-rheology properties of a fluid, only if the following conditions are fulfilled: (I) The fluid can be considered homogeneous at length scales of the order of the particle size or larger; (II) there are no specific (e.g. chemical) interactions between the particle and the fluid's components. The first restriction is fulfilled when the largest characteristic length scale of the fluid, such as the correlation length in a semi-dilute polymer solution or the mesh size in a gel network, is smaller than the particle's size. In the opposite situation, the fluid appears inhomogeneous. For example, a colloidal probe of size smaller than the mesh size of a gel experiences mostly the viscous environment of the solvent. This would lead to an apparent 'fluid-like' viscous behaviour, whereas the bulk properties would appear like those of a viscoelastic solid. The second requirement is necessary to ensure that the motion of the particle is only related to the dynamics of the fluid as a whole and not to local effects such as chemical bonds to the polymer chains or electrostatic interaction with them (Chen et al., 2003).

In the case of inhomogeneous media or non-ideal coupling with the fluid, the motion of a single particle does not reflect the bulk properties; however, it may be still useful for the investigation of local properties of the material. Nevertheless, there is a way to average the rheological properties of a fluid at length scales larger than the probe particle size. Indeed, the motion of a particle is hydrodynamically coupled to the motion of any other particle in the fluid. In a material the strain field generated by a particle entrains a second particle due to hydrodynamic interactions and the motions of both of them are correlated (Crocker et al., 2000; Levine and Lubensky, 2000). The correlation of the motion of two distinct particles contains information on the viscoelastic properties of the material averaged over the distance between them, excluding the effects of their local environments. In the so-called two-point correlation the following average is calculated (Crocker et al., 2000):

$$D_{rr}(R,\tau) = \left\langle \Delta r_R^i(t,\tau) \cdot \Delta r_R^j(t,\tau) \right\rangle_{t,i\neq j} \tag{8.23}$$

where D_{rr} is an averaged tensor product that contains the correlated motion of the particles and is a function of the separating distance R. The average is taken over all initial times and all pairs of distinct particles ($i \neq j$) at distance R. In a VPT measurement, R ranges from one particle distance to the lateral dimensions of the field of view. The subscript R in Eq. 8.23 denotes the projection of the particles' displacements along the vector connecting the two particles, for all lag-times τ. In the Laplace domain Eq. 8.23 becomes:

$$\tilde{D}_{rr}(R,s) = \frac{k_B T}{2\pi R s \tilde{G}(s)}. \tag{8.24}$$

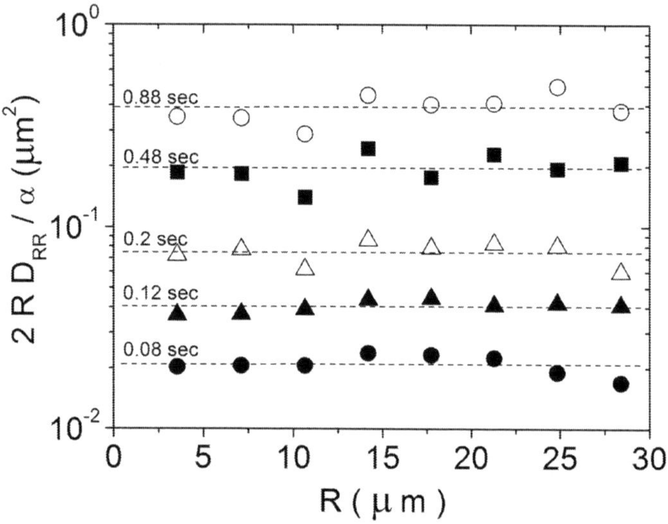

Figure 8.6 Test of the $1/R$ scaling of $D_{rr}(R, \tau)$; i.e. by plotting $\frac{2R}{a}D_{rr}(R, \tau)$ for different lag times. The horizontal dashed lines highlight the independence from R. Graph reproduced from (Papagiannopoulos et al., 2005a).

The above equation does not depend on the particle radius. This because it is not affected by the local environment and interactions. Moreover, by comparing Eq. 8.23 with Eq. 8.8 (i.e. the GSER) an effective MSD can be defined:

$$\left\langle \Delta r^2(\tau) \right\rangle_D = \frac{2R}{a} D_{rr}(R, \tau). \tag{8.25}$$

So that, the same data analysis used for the case of the single particle correlation can be used to extract the viscoelastic moduli. The subscript D in Eq. 8.25 indicates that the MSD has been obtained from the two-point correlation. It is important to mention that two-point correlation measurements require much longer recording times, e.g. 20 minutes or longer, compared to the one-point correlation, where a couple of minutes or so are normally enough. In Fig. 8.6 the $1/R$ scaling of $D_{rr}(R, \tau)$ confirms the correct two-point correlation measurement of long range hydrodynamic interactions in an incompressible fluid (i.e., aqueous solution of comb polyelectrolytes) for length scales as large as 60 particle radii. The resulting $\left\langle \Delta r^2(\tau) \right\rangle_D$ was found in good agreement with the

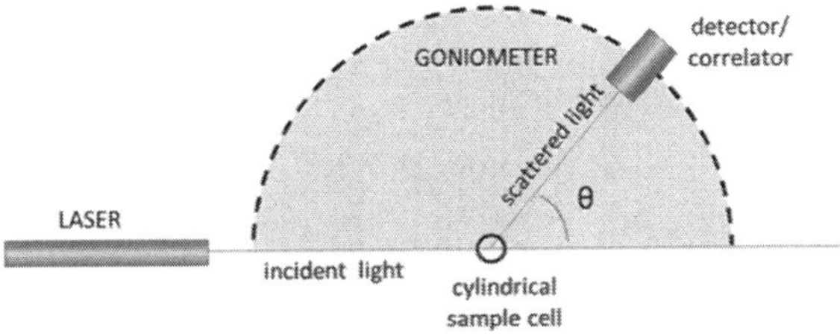

Figure 8.7 Schematic representation of a Dynamic Light Scattering (DLS) setup.

one-point MSD, and the validity of the microrheology results was confirmed by a direct comparison with conventional bulk-rheology measurements (Papagiannopoulos et al., 2005a).

8.4 Microrheology with Single Light Scattering

Dynamic Light Scattering (DLS) experiments measure the distribution of relaxation times in the scattered field auto-correlation function $g_1(\tau)$ from colloidal and macromolecular solutions (Berne and Pecora, 2000). The scattered intensity $I(t)$ is collected by a detector (Fig. 8.7) and a correlator evaluates the intensity auto-correlation function:

$$g_2(\tau) = \frac{\langle I(t) \cdot I(t+\tau) \rangle_t}{\langle I(t) \rangle_t \cdot \langle I(t+\tau) \rangle_t} \tag{8.26}$$

The electric field auto-correlation function ($g_1(\tau)$) is taken from the Siegert's relation:

$$g_2(\tau) = 1 + \beta g_1^2(\tau) \tag{8.27}$$

The field auto-correlation function, in the single scattering limit, from an ensemble of colloidal particles that undergo Brownian motion in solution is written as:

$$g_1(\tau) = \exp\left(-\frac{1}{6}q^2 \left\langle \Delta r^2(\tau) \right\rangle\right) \tag{8.28}$$

where q is the scattering wave vector, which is a function of the scattering angle θ, the wavelength of light λ and the solvent's refractive index n:

$$q = \frac{4\pi n}{\lambda} \sin\left(\frac{\theta}{2}\right) \tag{8.29}$$

The idea of single scattering microrheology is to measure the scattered intensity from probe particles dispersed in a complex fluid in order to extract its viscoelastic properties. The scattered intensity of the probes in solvent has to be significantly higher than the one from the underlying matrix, i.e. the complex fluid. In this way, the field auto-correlation function corresponds to the dynamics of the immersed colloidal particles (Dasgupta et al., 2002). The scattering from the probe particles is enhanced by increasing their concentration, but care must be taken so that the single-scattering assumption is still fulfilled. The time window explored by DLS experiments depends on the speed of the correlator. Typically, a fast correlator provides lag-times ranging from 10 ns to 1000 s. In practice, the longest time limit maybe compromised because of poor statistics of the auto-correlation function, especially in the case of weakly scattering samples. Of course, the quality of the statistics improves significantly with the increase of the total acquisition time. The longest useful lag-time is set by the point at which the correlation function drops to zero and it depends on the longest relaxation time of the complex fluid. The field auto-correlation function from diffusive particles drops to e^{-1} when $\frac{1}{6}q^2\left\langle\Delta r^2(\tau)\right\rangle = 1$; hence, by performing measurements at high q values (i.e. high scattering angles) one probes small particle displacements occurring at short lag-times (indeed, the MSD is normally an increasing function of time).

In general, microrheology measurements with DLS have a major limitation because of the multiple scattering effect. This undesired effect is commonly obtained when the concentration of probe particles is increased to overpass the scattering from the underlying fluid. However, we must mention that here exist two possible alternative methods that are based on the same principles of DLS measurements, but they work at different length scales and require different instrumentation. The first is based on X-Ray photon correlation spectroscopy. In particular, at small angles, multiple scattering effects are negligible and the probe particle concentration maybe increased according to the needs of the experiment (Papagiannopoulos et al., 2005b). Although this technique requires access to large scale facilities (such as Synchrotron radiation sources), it turns to be useful for experiments

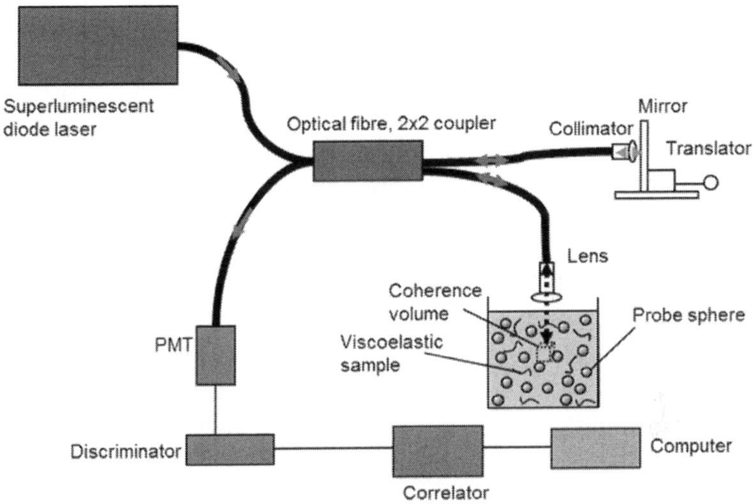

Figure 8.8 Experimental set-up of optical coherence tomography picorheology. Figure reproduced from (Sharma et al., 2008).

in highly opaque complex fluids, e.g. polymer melts and jammed colloids.

The second technique is based on an optical fiber method (Fig. 8.8). A superluminescent diode laser has a microscopic coherence length, which results into a coherence volumes of the order of some hundreds of pico-liters. The use of small coherence volumes suppresses multi scattering effects and also gives the possibility of local measurements, in a tomographic manner (Sharma et al., 2008). Translation of the mechanical stage attached to the mirror sets the path length within the sample to probe the local viscoelasticity across the sample. The back-scattered light collected by the optical fiber is mixed with the incident laser light (heterodyne mode) and collected by the photomultiplier tube (PMT).

8.5 Microrheology with Diffusing Wave Spectroscopy

In a dynamic light scattering experiment the multi-scattering regime is approached as the probe particle concentration is increased. At the extreme limit of turbid samples the propagation of light becomes diffusive and the electric field autocorrelation function needs a different interpretation. In diffusing wave spectroscopy (DWS), the laser light enters from one side of a rectangular sample cell (Fig. 8.9) and it is scattered by enough probe particles to follow a random walk before exiting the sample cell. In DWS measurements the probe particle concentration is of the order of $\sim$0.5% in volume; whereas in VPT it is of the order of $\sim$0.02%. In both the cases, these values of concentration are well below the critical concentration where particles start to interact with each other.

The electric field auto-correlation function in DWS (Pine et al., 1988) is the sum of the contributions of the individual paths that light can travel from the point of incidence to the point of detection. These contributions are weighted by the probability that a diffusing photon follows a given path. The electric field auto-correlation function is expressed as:

$$g_1(\tau) = \int_0^{+\infty} P(s) \cdot \exp\left(-\frac{1}{3}k_0^2 \left\langle \Delta r^2(\tau) \right\rangle \frac{s}{l*}\right) ds \tag{8.30}$$

where k_0 is the amplitude of the wave vector of the incident light, s is the path length, $\left\langle \Delta r^2(\tau) \right\rangle$ is the MSD of a microsphere probe and $l*$ is the light transport mean free path, which can be measured independently. Notice that, for the light propagation to be diffusive, $l*$ must be very small compared to the sample's length L. $P(s)$ is the path length distribution which depends on the geometry of the system. In transmission geometry (commonly used in DWS), the long path lengths contribute predominantly to the correlation function and therefore the measured MSDs are very small; thus very fast dynamics can be probed. In the transmission geometry, for a point source and for a sample of thickness L and infinite extent (this condition is accomplished by using rough glass for the sample-cell sides that are parallel to the incident beam) it has been shown (Weitz and Pine, 1993) that:

$$g_1(\tau) = C \int_Q^{+\infty} \exp\left(-\frac{\delta^2}{16}(\xi^2 - Q^2)\right) \cdot D(\xi, \epsilon, \zeta) d\xi \tag{8.31}$$

where d is the beam diameter, C is a normalisation constant such that $g_1(\tau \to 0) = 1, \delta = \frac{d}{L}, \epsilon = \frac{2l*}{3L}, \zeta = \frac{z_0}{L}$ and:

$$D(\xi, \epsilon, \zeta) = \frac{2\epsilon\left((1+\epsilon\xi) - (1-\epsilon\xi)e^{-2\zeta\xi}\right)}{(1+\epsilon\xi)^2 - (1-\epsilon\xi)^2 e^{-2\xi}} \tag{8.32}$$

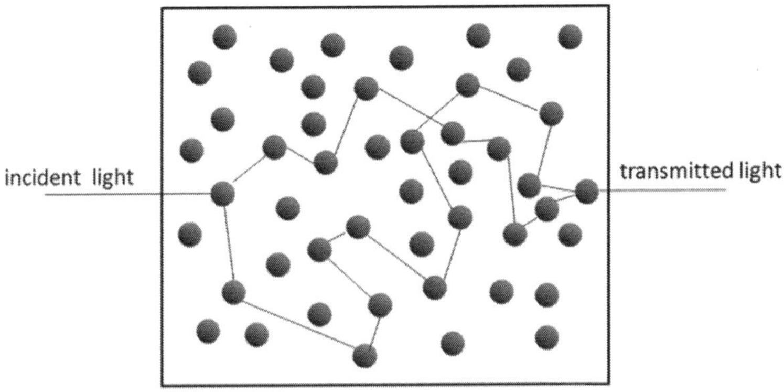

Figure 8.9 Diffusing propagation of light through a suspension of probe particles in the transmission geometry.

$$Q = \frac{L}{l*}\left(k_0^2\left\langle\Delta r^2(\tau)\right\rangle\right)^{1/2}. \tag{8.33}$$

The light source is point-like and it can be assumed that the transport of light becomes randomized at a distance $z_0 \approx l*$.

In order to apply Eqs. 8.30 and 8.31, the mean free path ($l*$) needs be known. This can be achieved by means of a suspension made of a Newtonian fluid of known viscosity (e.g. water) and with a known concentration of probe particles. This system will work as a reference to determine $l*$ in the complex fluid of unknown viscoelasticity. Indeed, in such a system, Eqs. 8.30 and 8.31 can be used with $l*$ as the only fitting parameter (Pine et al., 1988). Then, for measurements made with unknown complex fluids the same concentration of probes must be used and the transmission T through the sample must be also measured. Moreover, if the transmission in the reference liquid has also been measured, then $l*$ in the complex fluid is derived from the expression for diffusive transport with negligible absorption as $T \sim l*/(L + 4l*/3)$ (Weitz and Pine, 1993).

An example of combined study of DWS and VPT with an ultra-fast camera is shown in Fig. 8.10; where the viscoelasticity of the biological macromolecule aggrecan is probed in a wide time window, i.e. from 10^{-6} s to 10 s. The viscous-like behaviour at low frequencies ($G''(\omega) \propto \omega$ and $G'(\omega) \propto \omega^2$) is due to the entangled inter-diffusion of the agreecan macromolecules. At high

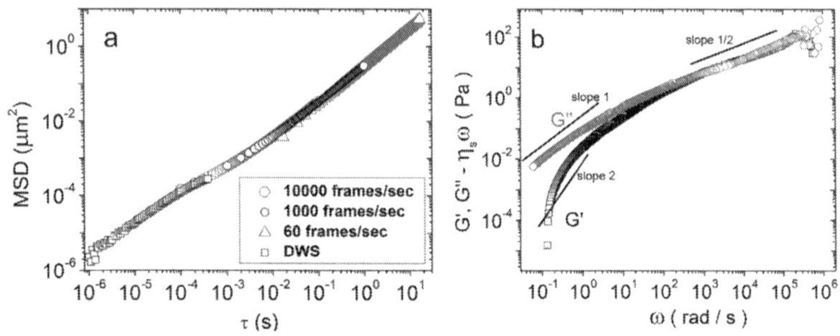

Figure 8.10 (a) MSD of amino-functionalized polystyrene microspheres (0.415 μm diameter) in monomeric aggrecan solution probed by VPT at several frame rates and DWS. (b) Resulting viscoelastic moduli evaluated over the entire frequency range obtained from both the experiment. Graphs reproduced from (Papagiannopoulos et al., 2008).

frequencies the Rouse model ($G'(\omega) \approx G''(\omega) \propto \omega^{1/2}$) reflects the internal dynamics of the agreecan macromolecules which is expected at high frequencies for linear polyelectrolytes in semi-dilute solutions (Papagiannopoulos et al., 2006). The ability of passive microrheology of probing a wide range of frequencies, practically inaccessible by conventional bulk-rheology methods, is one of the major advantages of these techniques. In particular, DWS is seen to be a powerful tool to probe fast dynamics.

Finally, it is worthy to remind that thanks to microrheology measurements with DWS, novel scaling behaviours have been observed for semi-flexible polymers (i.e., actin filament networks) at high frequencies (Palmer et al., 1999; Schnurr et al., 1997). In particular, the authors found that the complex shear modulus was proportional to $\omega^{3/4}$, which is in agreement with the theoretical prediction (Morse, 1998) for semi-flexible polymers at high frequencies.

8.6 Microrheology with Magnetic Tweezers

All the microrheology methods presented so far are based on the analysis of thermally driven random motion of micron sized probe particles. Thus no external force fields are required to induce the particle displacement; reason for which they are classified as *passive microrheology* techniques. This means that the measurements performed with such techniques fall inherently in the material's linear viscoelastic regime. This is a great advantage over traditional bulk-rheology methods, where the range of linearity must always be determined before performing the measurement. However, passive microrheology methods are limited (at the top-end) on the range of viscosities that could be measured. This because as the fluid viscosity increases the thermal fluctuations (i.e. the MSD) decreases down to a point that they becomes too small to be detected by the experimental apparatus. In order to overcome this issue, *active microrheology* techniques have proved to be a useful complement or substitute to passive microrheology methods in the case of high viscous fluids or gel-like samples. In particular, in active microrheology, a time-dependent external force field is applied to induce the probe displacement and the viscoelastic nature of the fluid lies in the relationship between these two quantities; as for conventional rheology experiment where a sinusoidal stress is applied and the resulting material deformation is measured.

Let us consider a simple magnetic microrheometer configuration, like the one used in (Tassieri et al., 2010b) and here schematically shown in Fig. 8.11. In this case, the magnetic field is generated by two electromagnetic coils, and a magnetic moment ($\vec{m}(t)$) is induced on the superparamagnetic beads (of few microns in diameter) suspended in the fluid of interest. The force ($F(t)$) exerted on a superparamagnetic particle inside a magnetic field ($\vec{B}(t)$) is proportional to the square of the electric current ($I(t)$) flowing into the coils:

$$\vec{F}(t) \propto \nabla B^2(t) \propto I^2(t) \tag{8.34}$$

this because for superparamagnetic materials $\vec{m}(t) \propto \vec{B}(t)$ and for a solenoid $|\vec{B}(t)| \propto I(t)$. To control the applied force it is necessary to find the exact proportionality in Eq. 8.34 between the force and the current. This is possible by analysing the particle motion in a purely viscous medium (e.g., a glycerol-water mixture) of known viscosity (η) combined with the Stoke's equation that correlates the particle velocity ($v(t)$) to the friction force (f_v). The latter equals instant by instant the induced magnetic force:

$$\vec{F}(t) \equiv \vec{f}_v(t) = 6\pi\eta a\vec{v}(t) \tag{8.35}$$

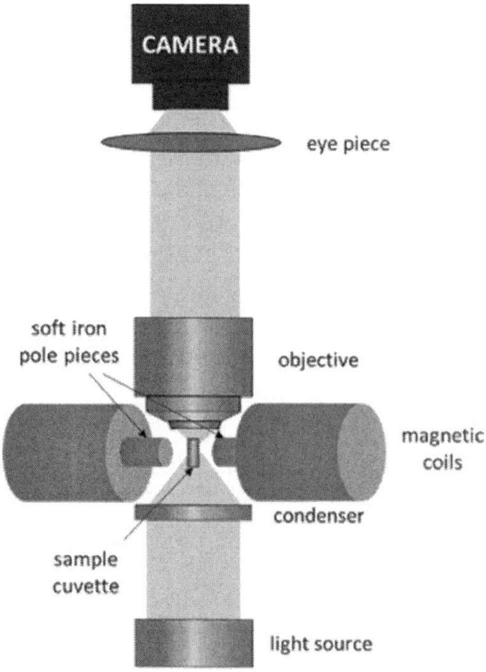

Figure 8.11 Schematic representation of a generic magnetic tweezers setup. A time-dependent magnetic field is generated by two electromagnets. The interaction between the gradient of the magnetic field and the (induced) magnetic dipole of the magnetic probe results in a net magnetic force acting on the probe in the direction of the maximum gradient.

Therefore, by controlling the coils' current and measuring the particle displacement (thus the velocity) a calibration curve $F(t)$ *vs.* $I(t)$ can be drawn.

Once the magnetic microrheometer is calibrated, two *modus operandi* are possible for microrheological purposes: I) An oscillatory measurement or II) a creep measurement. In the first case a sinusoidal force is applied $F(t) = F_0 \sin(\omega t)$ (where F_0 is the force amplitude) and the resulting sinusoidal bead displacement ($x(t) = x_0(\omega)\sin(\omega t + \varphi(\omega))$) is measured; where $x_0(\omega)$ is the frequency-dependent amplitude of the bead displacement and $\varphi(\omega)$ is the phase-shift between the force and the bead displacement. In this case, the linear viscoelastic properties of the fluids can be determined by means of the following two equations:

$$G'(\omega) = \frac{F_0}{6\pi\eta a x_0(\omega)}\cos(\varphi(\omega)) \tag{8.36}$$

$$G''(\omega) = \frac{F_0}{6\pi\eta a x_0(\omega)}\sin(\varphi(\omega)) \tag{8.37}$$

where $G'(\omega)$ and $G''(\omega)$ are the elastic and the viscous moduli of the fluid, respectively.

In the second case, a creep measurement is achieved by applying a constant force (F_0) and monitoring the bead displacement $(x(t))$; similarly to the bulk-rheology measurement where a constant stress is applied instantaneously and the material strain (i.e., deformation) is monitored. In the case of a microsphere of radius a suspended into a fluid, the creep compliance $(J(t))$ is defined as it follows:

$$J(t) = 6\pi a\frac{x(t)}{F_0}. \tag{8.38}$$

The frequency-dependent fluid's LVE properties are then educed by means of Eq. 6.12 and the Fourier transform of the creep compliance:

$$G^*(\omega) = \frac{1}{i\omega\hat{J}(\omega)} \equiv \frac{F_0}{6\pi a}\frac{1}{i\omega\hat{x}(\omega)} \tag{8.39}$$

where $\hat{x}(\omega)$ is the Fourier transform of the bead displacement $x(t)$.

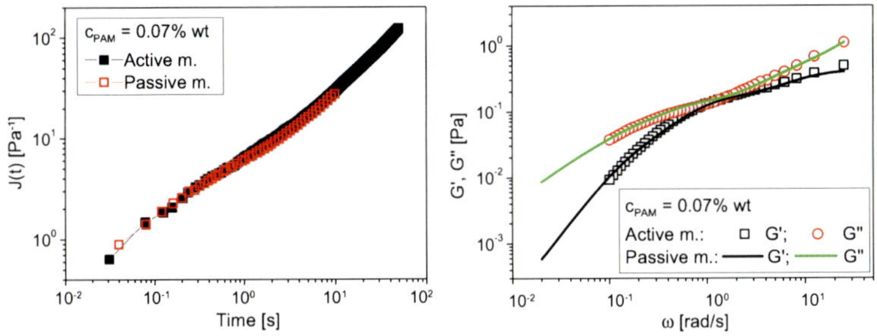

Figure 8.12 (Left) Active and Passive microrheology measurements of creep compliance *versus* time for a polyacrylamide (PAM) solution at a concentration of 0.07% *wt.* (Right) Comparison between the loss and storage moduli *versus* frequency obtained from the conversion of passive and active microrheology measurements of the creep compliances shown on the left. The graphs have been extracted from (Tassieri, 2007).

Figure 8.12 shows both a typical result of a creep measurement performed with a magnetic rheometer, but also a direct comparison with a passive VPT microrheology measurement performed on the same fluid; the agreement is apparent, with the added advantage of exploring a wider range of frequencies in the case of active microrheology.

Chapter 9

Microrheology with Optical Tweezers*

Manlio Tassieri
Division of Biomedical Engineering, School of Engineering, University of Glasgow, Glasgow G12 8LT, UK
manlio.tassieri@glasgow.ac.uk

9.1 Introduction

Since their first appearance in the 1970s (Ashkin, 1970), optical tweezers (OT) have been extensively developed and have proved to be an invaluable tool for a variety of applications throughout the biophysical sciences (Ashkin and Dziedzic, 1987; Block et al., 1989; Finer et al., 1994; Mehta et al., 1999; Meyer et al., 2006; Tskhovrebova et al., 1997; Yoon et al., 2008). This success relies essentially on the fact that OT can be considered as exceptionally sensitive mechanical transducers able to resolve pN forces and nm displacements, with high temporal resolution (down to μsec); offering to a broad scientific community the opportunity of mechanically sensing the systems' dynamics at relatively small length scales, i.e., down to a molecular level. Hence, the rheological interest of OT

*Much of the material in this chapter has appeared in (Preece et al., 2011; Tassieri et al., 2010a, 2012).

Microrheology with Optical Tweezers: Principles and Applications
Edited by Manlio Tassieri

ISBN 978-981-4669-18-4 (Hardcover), 978-981-4669-19-1 (eBook)
www.panstanford.com

for inverting the classical top-bottom approach of bulk-rheology measurements addressed to investigate the molecular dynamics. Indeed, with the advent of OT, rheologists have the opportunity of a new mechanical tool for measuring single molecular dynamics and developing bottom-up rheological models aimed at predicting the bulk-flow of matter.

The physics underpinning the working principles of the OT relies on the ability of a focused laser beam to trap, in three dimensions, micron-sized dielectric particles suspended in a fluid as briefly described here and in details in Chapter 4, but also in (Ashkin, 1970; Ashkin and Dziedzic, 1971; Ashkin et al., 1986). The trapping process is achieved by optically guiding a (monochromatic) laser beam through a high magnification objective, with high numerical aperture, which also allows visualisation of the micro-environment surrounding the trapped probe. The trapping force is generated by the combined action of both the gradient of the laser intensity profile and the difference between the refractive indexes of the materials constituting the fluid and the probe. OT are very often built around optical microscopes, like those commonly used for biological studies, and are equipped with a fast probe-position detector. The latter is usually chosen to be either a charge-coupled device (CCD) camera (Gibson et al., 2008) or a quadrant photodiode (QPD) (Gittes and Schmidt, 1998); both of which provide the tracer trajectory with a spatial resolution of the order of nm (Keen et al., 2007), but the first has a detection rate upper limit of the order of kHz, while the second can reach rates of the order of MHz. Accessing the time-dependent trajectory of a micron-sphere, to high spatial and temporal resolution, is one of the basic principles behind microrheology techniques as described in (Mason and Weitz, 1995; Squires and Mason, 2010) and in this book.

9.2 Optical Tweezers Calibration

As for any mechanical transducer also optical tweezers require to be calibrated, both in space and in strength, as already discussed in Chapter 5 for the case when the trapped particle is suspended in a Newtonian fluid. However, different is the story when the fluid

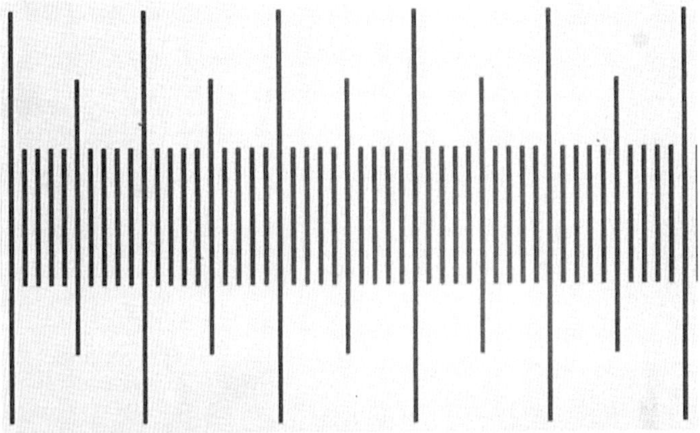

Figure 9.1 A typical image of a microscope calibration scale bar. The distance between two consecutive bars is 10 μm.

is non-Newtonian and its elasticity may affect the calibration of the trap stiffness, as explained hereafter.

9.2.1 *Spatial Calibration*

In the case of OT equipped with a CCD camera, the spatial calibration refers to the process of correlating the pixel's size of an acquired image to the real dimensions of the features captured by the image. This process can be easily performed by means of a calibration slide on which a metric scale bar is drawn, as shown in Fig. 9.1. The average value of repeated measurements of the distance between two bars provides an accurate estimation of the pixel/unit conversion factor. In the case of OT equipped with a QPD device, the spatial calibration is not straightforward as for the previous case and it is still object of studies, reason for which we refer the reader to dedicated works such as (Keen et al., 2007; Tolic-Norrelykke et al., 2006).

Once the probe-position detector is calibrated, then in order to measure a particle trajectory, an image analysis process is required. In the majority of the cases, home made routines are developed by the OT operators using a variety of image analysis softwares. However, the basic principles underpinning the greater part of all

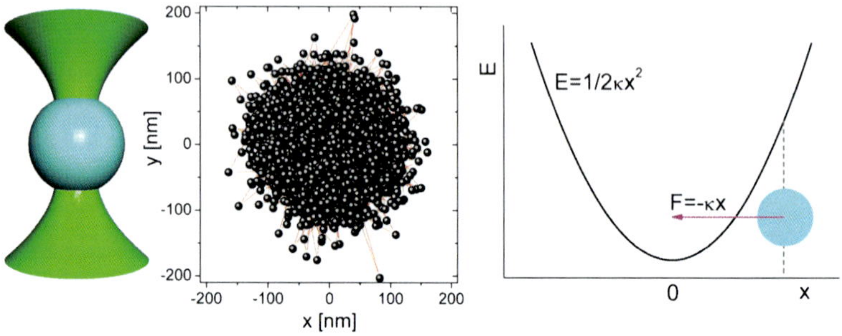

Figure 9.2 (Left) A schematic drawing of an optically trapped micro-sphere. (Center) The trajectory made of 10^6 points of a 5 μm diameter glass bead and acquired at 1 kHz. (Right) A schematic drawing of the force exerted on a micro-sphere because of the harmonic potential generated by the optical tweezers.

particle tracking procedures have been clearly explained by John C. Crocker and David G. Grier in their pioneering work (Crocker and Grier, 1996).

9.2.2 *Elastic Constant Calibration*

The most important feature of optical tweezers is that the confining action exerted by the highly focussed laser beam on a spherical particle can be modelled as a harmonic potential E of force constant κ (as schematically shown in Fig. 9.2):

$$E = \frac{1}{2}\kappa\langle r^2\rangle_{\text{eq}} \tag{9.1}$$

where $\langle r^2\rangle_{\text{eq}}$ is the time-independent variance at equilibrium of the particle position from the trap centre, the origin of $\vec{r}$. Consequently, the restoring force exerted on the particle is simply given by the negative gradient of E:

$$\vec{F} = -\nabla E = -\kappa\vec{r} \tag{9.2}$$

In order to quantify the force exerted on the particle, a calibration of the elastic constant (trap stiffness) κ is required. This is simply achievable by appealing to the Principle of Equipartition of Energy:

$$\frac{d}{2}k_{\text{B}}T = \frac{1}{2}\kappa\langle r_j^2\rangle_{\text{eq}} \tag{9.3}$$

where k_B is Boltzmann's constant, T is absolute temperature and $\langle r_j^2 \rangle$ is the time-independent variance of the Cartesian component ($j = x, y, z$) of the d-dimentional vector describing the particle's displacement from the trap center, $\vec{r} = \vec{0}$. It is important to note that, for non-symmetric traps (i.e., $\kappa \neq \kappa_j, \forall j$) Eq. 9.3 is still valid but in one dimension, with κ replaced by κ_j evaluated for each component:

$$k_B T = \kappa_j \langle r_j^2 \rangle_{\text{eq}} \tag{9.4}$$

Despite the great variety of methods for determining the trap stiffness (e.g., using the drag force or the power spectrum, as described in (Berg-Sørensen and Flyvbjerg, 2004; Jun et al., 2014; Neuman and Block, 2004; Tolic-Norrelykke et al., 2006)), Eq. 9.3 (or 9.4) provides the *only* such measurement that is independent of the viscoelastic properties of the fluid under investigation and is thus essential for proper calibration. This is because, whatever the elasticity of the unknown fluid, its contribution to the time-independent constraining force must vanish at long times (because at rest the fluid's elastic shear modulus goes to zero as the time goes to infinity). Thus the trap stiffness is easily determined by means of Eq. 9.3 applied to a *sufficiently long measurement* via the evaluation of the time-independent variance of the confined particle position.

9.3 Microrheology with Static Optical Tweezers

9.3.1 *Solving a Generalised Langevin Equation for Static OT*

When a micron-sized spherical particle is suspended in a fluid at thermal equilibrium, it experiences random forces leading to Brownian motion, driven by the thermal fluctuations of the fluid's molecules. Analysis of the particle's trajectory reveals information on the viscoelastic properties of the suspending fluid, as demonstrated in the pioneering work of (Mason and Weitz, 1995) that established the field of microrheology.

Similarly, when the thermal fluctuations of a probe are constrained by a stationary harmonic potential generated by optical tweezers, the analysis of the particle's trajectory has the potential of revealing information on both the viscoelastic nature of the suspending fluid and the confining force of the trap. The random motion of the trapped particle could be described by means of a generalised Langevin equation almost identical to that used by Mason and Weitz, but with an added term that takes in account the OT restoring force (i.e., Eq. 9.2):

$$m\vec{a}(t) = \vec{f}_R(t) - \int_0^t \zeta(t-\tau)\vec{v}(\tau)d\tau - \kappa\vec{r}(t). \tag{9.5}$$

where m is the mass of the particle, $\vec{a}(t)$ is its acceleration, $\vec{v}(t)$ its velocity and $\vec{f}_R(t)$ is the usual Gaussian white noise term, modelling stochastic thermal forces acting on the particle. The integral term, which incorporates a generalised time-dependent memory function $\zeta(t)$, represents viscous damping by the fluid. Following the same assumptions made by Mason and Weitz for the case of freely diffusing particles, i.e., that the bulk Laplace-frequency-dependent viscosity of the fluid $\tilde{\eta}(s)$ is proportional to the microscopic memory function:

$$\tilde{\zeta}(s) = 6\pi a\tilde{\eta}(s) \tag{9.6}$$

where a is the bead radius, Eq. 9.5 can be solved in terms of either the normalised mean-square displacement (NMSD) (Tassieri et al., 2010a):

$$\Pi(\tau) = \frac{\langle \Delta r^2(\tau)\rangle_{t_0}}{2\langle r^2\rangle_{t_0}} \equiv \frac{\langle [(r(t_0+\tau) - r(t_0)]^2\rangle_{t_0}}{2\langle r^2\rangle_{t_0}} \tag{9.7}$$

or the normalised position autocorrelation function (NPAF) (Preece et al., 2011):

$$A(\tau) = \frac{\langle \vec{r}(t_0)\vec{r}(\tau)\rangle_{t_0}}{\langle r^2\rangle_{t_0}} \tag{9.8}$$

which, by time-translation invariance, are both functions only of the time interval (lag-time) $\tau = \Delta t = (t - t_0)$. Moreover, the brackets $\langle \cdots \rangle_{t_0}$ denote an average over all initial times t_0. Figure 9.3 shows both the above time-averaged quantities evaluated from a typical trajectory of an optically trapped 5 μm bead suspended in water.

Let us show step by step how Eq. 9.5 can be solved in terms of the NPAF and for simplicity let us rewrite it without the vectorial notation:

$$m\ddot{r}(t) = f_R(t) - \int_0^t \zeta(t-\tau)\dot{r}(\tau)d\tau - \kappa r(t) \tag{9.9}$$

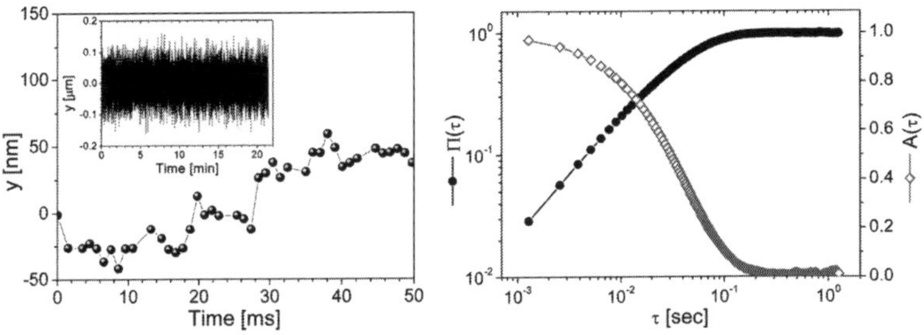

Figure 9.3 (Left) The y-component of the trajectory of an optically trapped 5 µm bead suspended in water over a period of 50 ms. The inset shows the same component as before, but over the entire experiment of 22 min. (Right) The normalised position autocorrelation function $A(\tau)$ and the normalised mean square displacement $\Pi(\tau)$ *versus* lag-time τ evaluated from the trajectory shown on the left.

where $\dot{r}(t)$ and $\ddot{r}(t)$ denote respectively the first and the second time derivative of the particle position $r(t)$. The first step is to perform the Laplace transform of Eq. 9.9:

$$m[s^2\tilde{r}(s) - sr(t_0) - \dot{r}(t_0)] = \tilde{f}_R(s) - \tilde{\zeta}(s)[s\tilde{r}(s) - r(t_0)] - \kappa\tilde{r}(s) \tag{9.10}$$

where s is the Laplace frequency and the convolution integral is transformed into the product of the two Laplace-transformed integrand functions. The following step is to multiply both the sides of Eq. 9.10 by the term $r(t_0)$ (which is a constant in the Laplace domain) and then take the average over all initial times t_0:

$$\begin{aligned} m[s^2\langle r(t_0)\tilde{r}(s)\rangle_{t_0} - s\left\langle r^2(t_0)\right\rangle_{t_0} - \langle r(t_0)\dot{r}(t_0)\rangle_{t_0}] = \\ \left\langle r(t_0)\tilde{f}_R(s)\right\rangle_{t_0} - \tilde{\zeta}(s)[s\langle r(t_0)\tilde{r}(s)\rangle_{t_0} - \left\langle r^2(t_0)\right\rangle_{t_0}] \\ - \kappa\langle r(t_0)\tilde{r}(s)\rangle_{t_0} \end{aligned} \tag{9.11}$$

where the terms $\langle r(t_0)\dot{r}(t_0)\rangle_{t_0}$ and $\left\langle r(t_0)\tilde{f}_R(s)\right\rangle_{t_0}$ are both equal to zero because for a stochastic process the averaged functions are uncorrelated. Therefore, Eq. 9.11 becomes:

$$\begin{aligned} ms[s\langle r(t_0)\tilde{r}(s)\rangle_{t_0} - \left\langle r^2(t_0)\right\rangle_{t_0}] = \\ - \tilde{\zeta}(s)[s\langle r(t_0)\tilde{r}(s)\rangle_{t_0} - \left\langle r^2(t_0)\right\rangle_{t_0}] - \kappa\langle r(t_0)\tilde{r}(s)\rangle_{t_0} \end{aligned} \tag{9.12}$$

where the term $\langle r(t_0)\tilde{r}(s)\rangle_{t_0}$ is the Laplace transform of the particle position autocorrelations function (PAF) and $\langle r^2(t_0)\rangle_{t_0}$ is the time-independent variance of the particle position. From Eqs. 9.12 and 9.8 we can write:

$$ms[s\tilde{A}(s) - 1] = -\tilde{\xi}(s)[s\tilde{A}(s) - 1] - \kappa\tilde{A}(s) \tag{9.13}$$

and

$$ms = -\tilde{\xi}(s) - \kappa\frac{\tilde{A}(s)}{s\tilde{A}(s) - 1} \tag{9.14}$$

from which

$$\tilde{\xi}(s) = \kappa\frac{\tilde{A}(s)}{1 - s\tilde{A}(s)} - ms. \tag{9.15}$$

By combining Eqs. 9.6 and 9.15 we can write:

$$\tilde{\eta}(s) = \frac{\kappa}{6\pi a}\left(\frac{\tilde{A}(s)}{1 - s\tilde{A}(s)} - \frac{ms}{\kappa}\right). \tag{9.16}$$

Moreover, given that

$$G^*(\omega) \equiv s\tilde{\eta}(s)\big|_{s=i\omega}, \tag{9.17}$$

the complex modulus $G^*(\omega)$ can be expressed directly in terms of the time-dependent NPAF:

$$G^*(\omega) = \frac{\kappa}{6\pi a}\left(\frac{i\omega\hat{A}(\omega)}{1 - i\omega\hat{A}(\omega)} + \frac{m\omega^2}{\kappa}\right) \tag{9.18}$$

where $\hat{A}(\omega)$ is the Fourier transform of $A(\tau)$.

As anticipated, it is possible to express the fluid's complex modulus as a function of the NMSD. Indeed, with reference to Chapter 7, Eq. 9.16 is actually equivalent to Eq. 7.155, which can be reorganised to provide an alternative expression to Eq. 9.18:

$$G^*(\omega) = \frac{\kappa}{6\pi a}\left(\frac{2\langle r^2\rangle_{t_0}}{i\omega\left\langle \widehat{\Delta r^2}(\omega)\right\rangle_{t_0}} + \frac{m\omega^2}{\kappa} - 1\right). \tag{9.19}$$

It is important to note that, for micron sized particles, the term $m\omega^2/\kappa$ that appears in both Eqs. 9.18 and 9.19 is negligible for frequencies values $\omega \ll$ MHz. Therefore, by considering Eqs. 9.7, 9.18 and 9.19 it is possible to write:

$$G^*(\omega)\frac{6\pi a}{\kappa} = \frac{i\omega\hat{A}(\omega)}{1 - i\omega\hat{A}(\omega)} \equiv \left(\frac{1}{i\omega\hat{A}(\omega)} - 1\right)^{-1}$$
$$\equiv \left(\frac{1}{i\omega\hat{\Pi}(\omega)} - 1\right) \equiv \frac{\hat{A}(\omega)}{\hat{\Pi}(\omega)} \quad (9.20)$$

where $\hat{\Pi}(\omega)$ is the Fourier transform of $\Pi(\tau)$. This is possible because the quantities $A(\tau)$ and $\Pi(\tau)$ are simply related to each other (Preece et al., 2011), both in the time- and in the frequency-domain. Indeed, by expanding the numerator of Eq. 9.7 one obtains:

$$\Pi(\tau) = \frac{\left\langle r^2(t_0+\tau)\right\rangle_{t_0} + \left\langle r^2(t_0)\right\rangle_{t_0} - 2\left\langle \vec{r}(t_0)\vec{r}(\tau)\right\rangle_{t_0}}{2\left\langle r^2\right\rangle_{t_0}} \equiv 1 - A(\tau) \quad (9.21)$$

where $\left\langle r^2(t_0+\tau)\right\rangle_{t_0} \equiv \left\langle r^2(t_0)\right\rangle_{t_0}$ for *sufficiently long* measurements. In addition, the Fourier transform of Eq. 9.21 gives:

$$i\omega\hat{\Pi}(\omega) = 1 - i\omega\hat{A}(\omega) \quad (9.22)$$

thus the equalities in Eq. 9.20.

In principle, Eqs. 9.18, 9.19 and 9.20 are simple expressions relating the material's complex shear modulus $G^*(\omega)$ to the observed time-dependent bead trajectory $\vec{r}(t)$ *via* the Fourier transform of one of the related time-averaged quantities. In practice, the evaluation of these Fourier transforms, given only a finite set of data points over a finite time domain, is non-trivial since interpolation and extrapolation from those data can yield artifacts that lie within the bandwidth of interest.

The first attempt to address this issue for microrheology (Mason and Weitz, 1995) was to fit the measured $\tilde{G}(s)$ to a preconceived functional form, and then use analytic continuation (swapping Laplace for Fourier frequency, $s \to i\omega$) to recover $G'(\omega)$ and $G''(\omega)$. A later approximate method (Mason, 2000) recovered the moduli from the log slope of the MSD, without numerical inversions, and was subsequently improved upon (Dasgupta et al., 2002) by accounting for curvature.

An alternative method has been proposed by (Evans et al., 2009), to convert creep compliance $J(t)$ into $G'(\omega)$ and $G''(\omega)$ directly, without transforms or fitting functions. This method is based on the interpolation of the finite data set by means of a piecewise-linear function. The general validity of the proposed procedure makes it equally applicable to find the Fourier transform $\hat{g}(\omega)$ of any time-dependent function $g(t)$ that vanishes for negative t, sampled at a finite set of data points (t_k, g_k), where $k = 1 \ldots N$, which extend over a finite range, and *need not* be equally spaced:

$$-\omega^2 \hat{g}(\omega) = i\omega g(0) + \left(1 - e^{-i\omega t_1}\right) \frac{(g_1 - g(0))}{t_1} + \dot{g}_\infty e^{-i\omega t_N} + \sum_{k=2}^{N} \left(\frac{g_k - g_{k-1}}{t_k - t_{k-1}}\right) \left(e^{-i\omega t_{k-1}} - e^{-i\omega t_k}\right) \quad (9.23)$$

where $\dot{g}_\infty$ is the gradient of $g(t)$ extrapolated to infinite time and $g(0)$ is the value of $g(t)$ extrapolated to $t = 0$ from above. The method proposed by (Evans et al., 2009) has been then improved by (Tassieri et al., 2012) as described hereafter and in Appendix.

9.3.2 *Data Analysis*

Although Eq. 9.23 has been successfully employed in both classical bulk rheology (Evans, 2009; Spruijt, 2010) and microrheology studies (Preece et al., 2011; Tassieri et al., 2010a,b; Yanagishima et al., 2011), when it is applied to microrheology measurements performed with optical tweezers, there remain *two* issues, common to all the other methods described above, that compromise the quality of the results. Nevertheless, the data analysis procedure introduced by (Tassieri et al., 2012) markedly reduces the undesired effects of these experimental issues, providing a useful tool for microrheology.

9.3.2.1 Interpolation artefacts

The first issue is related to the Nyquist-Shannon sampling theorem (Shannon, 1949):

> *If a function contains no frequencies higher than W cps, it is completely determined by giving its ordinates at a series of points spaced* $1/(2W)$ *seconds apart.*

Conversely, one cannot expect to recover complete information about the fluid's LVE properties from the Fourier transform of a discrete data set (e.g., MSD, NMSD or NPAF) at frequencies higher than one half of the data acquisition rate (AR) if the rate is uniform. Above that frequency, the information in the Fourier transform is artificial, describing the details of the interpolation scheme between the data points. Unfortunately, those artifacts will almost certainly spread some way *below* the Nyquist limit, contaminating the experimental results. This occurs due to the broad spectrum of the sharp corners between straight-line segments in the interpolation scheme used to derive Eq. 9.23 (see Appendix).

This first issue is simply resolved by *virtually* oversampling the time-averaged functions (e.g., $\Pi(\tau)$) that contain the fluid's LVE properties. Oversampling is a very common procedure in signal processing and it consists of sampling a signal with a sampling frequency f_s much higher than the Nyquist rate $2B$, where B is the highest frequency contained in the original signal. A signal is said to be oversampled by a factor of $\beta \equiv f_s/(2B)$ (Lathi, 2004). Microrheological measurements performed with optical tweezers have a maximum value of B limited by the detector's acquisition rate (AR). Therefore, the oversampling process can be achieved by interpolating the measured time-averaged functions (e.g., $A(\tau)$) with a natural cubic spline, which is piecewise cubic and twice continuously differentiable.

Having generated the larger oversampled data set from the original data set, it remains important to use Eq. 9.23 to find its Fourier transform, since that equation correctly assumes causality (the function vanishes for negative time), extrapolates to infinite time (using the parameter $\dot{g}_\infty$) and guarantees that no other assumptions bias the data. In contrast to the cubic-spline interpolation, a fitting procedure based on preconceived models would *doctor* the experimental results; the interpolation procedure preserves the original data within the bandwidth limits imposed by the sample rate. It is important to note that, following the oversampling and conversion, the resulting rheological data (i.e., the frequency-dependent viscoelastic moduli) should only be examined at frequencies within the experimentally valid frequency window, since no real information exists above the Nyquist frequency. This data-analysis procedure yields artifact-free results right up to the Nyquist limit.

A straightforward validation of the solution to this first issue is achievable by apply it to a simple (analytical) expression of NPAF, i.e., a single exponential decay, as if it were obtained from an infinitely long measurement of the thermal fluctuations of a particle harmonically trapped in a Newtonian fluid:

$$A(\tau) = e^{-\lambda\tau} \tag{9.24}$$

where $\lambda = \kappa/(6\pi a\eta)$, is the relaxation rate of the compound system (OT, bead, fluid), known as *corner frequency* when the thermal fluctuations of an optically trapped bead are analysed in terms of the power spectral density (Berg-Sørensen and Flyvbjerg, 2004). *Coincidentally*, Eq. 9.24 is the dynamic response of an ideal Kelvin-Voigt material (see Chapter 6), for which the viscoelastic frequency spectrum is known to have a simple analytical form:

$$G^*_{\text{tot}}(\omega) = \kappa/(6\pi a) + i\eta\omega \tag{9.25}$$

or equivalently

$$\bar{G}^*_{tot}(\omega) = 1 + i\bar{\omega} \tag{9.26}$$

where $\bar{G}^*_{\text{tot}}(\omega) \equiv (6\pi a/\kappa)\, G^*_{\text{tot}}(\omega)$ and $\bar{\omega} = \omega/\lambda$.

Figure 9.4(left) shows an example in which the idealised NPAF has been sampled at a frequency of $f_s = 1$ kHz, as if it were obtained from thermal fluctuations of a 5 μm diameter sphere constrained by an OT with trap stiffness of $\kappa = 1$ μN/m and suspended in a Newtonian fluid of viscosity $\eta = 0.896$ mPa·s, where the bead trajectory had been acquired at $AR \equiv f_s$ and for an infinite time, up to a lag-time of 1*s*. The dynamic response of the system can be obtained by Fourier transforming $A(\tau)$ *via* Eq. 9.23; the results are shown in Fig. 9.4(right). It is clear that artefacts in the frequency domain, where $\bar{G}'_{\text{tot}}(\omega)$ starts to diverge from its expected value (due to the finite sampling rate and implicit piece-wise-linear interpolation in the time domain), begin at $\omega \simeq \lambda$ (i.e., $\bar{\omega} \simeq 1$), whereas, $\bar{G}''_{\text{tot}}(\omega)$ starts to deviate from its expected value (i.e., $\bar{\omega}$) only at $\omega \simeq \lambda^2$.

Figure 9.5 demonstrates that the correct values of the LVE moduli are recovered by oversampling the discrete data shown

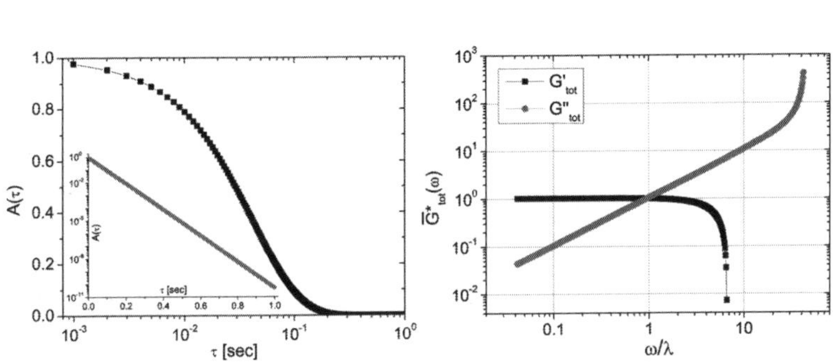

Figure 9.4 (Left) The normalised position autocorrelation function $\left(A(\tau) = e^{-\lambda\tau}\right)$ *vs.* lag-time; where $\lambda = \kappa/(6\pi a\eta) \simeq 24s^{-1}$. The NPAF has been built using the following parameters' values: $f_s = 1$ kHz, $\kappa = 1$ μN/m, $a = 2.5$ μm, $\eta = 0.896$ mPa·s. The inset shows the same data as above, but with the semi-log representation of the axis inverted. (Right) The normalised complex modulus *vs.* frequency evaluated *via* Eq. 9.20 and by means of Eq. 9.23 applied to the data shown on the left.

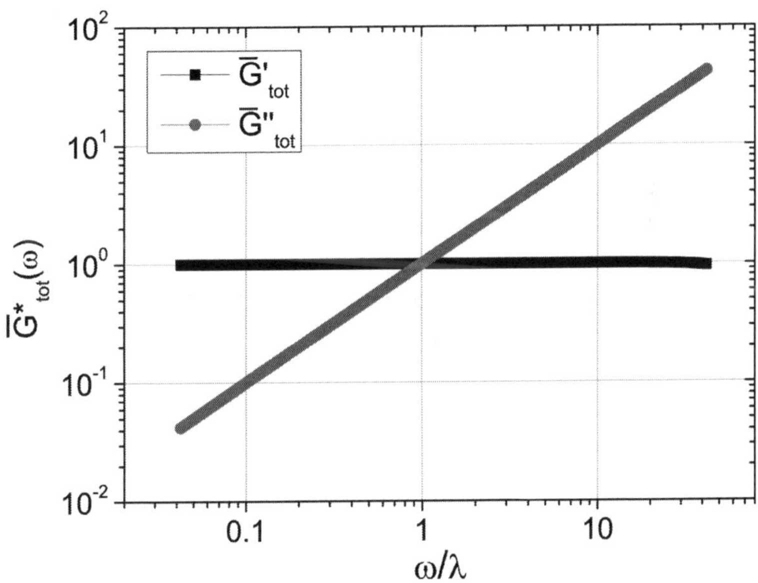

Figure 9.5 The normalised complex modulus *vs.* frequency evaluated *via* Eq. 9.20 and by means of Eq. 9.23 applied to the data shown in Fig. 9.5(left), but interpolated with a natural cubic spline function having $f_s \cong 8.2$ MHz. This correct normalisation confirms the validity of the data analysis method.

in Fig. 9.4(left) to a sufficiently high value of $\beta \equiv f_s/(2AR) \gg \omega/AR$, using a natural cubic spline interpolation; i.e., $f_s \cong 8.2$ MHz (implying $\beta \cong 4100$). In Figure 9.5, both the moduli now show the expected values (i.e., Eq. 9.26) over the *entire* range of explored frequencies.

9.3.2.2 Noise

The second issue relates to the accuracy with which the data (i.e., $\langle \Delta r^2(\tau) \rangle$, $\Pi(\tau)$ or $A(\tau)$) are evaluated, especially at long lag-times. Indeed, since both the functions $\Pi(\tau)$ and $A(\tau)$ (and those related such as the variance) are time-averaged quantities, they become *exact* only in the limit of infinite measuring time (or equivalently for $N \to \infty$), which is unachievable in reality.

In real experiments this uncertainty affects the results in the following way. Since the Fourier transform is an integral operator, all the short time-scale deviations of the measured time-averaged function from its expected value will communally contribute noise to the results at high frequencies, especially those occurring at long lag-times where the averages are inexorably less accurate. This is clearly shown in Fig. 9.6(left), where the agreement between the measured NMSD and its prediction, *via* Eqs. 9.21 and 9.24, is very good at small lag-times; whereas it becomes worse at long lag-times (see inset of Fig. 9.6(left)). Although *all* of the data contribute to the high-frequency *noise*, the high-frequency *signal* derives predominantly from the measurements at short lag-times. In long-duration experiments, this signal can become swamped by noise from the large quantity of long lag-time data. A simple but crude solution to this problem, to improve the signal-to-noise ratio, is obtained by reducing the data density (i.e., the sampling rate) at long lag-times. This is achieved by evaluating the time-averaged functions (e.g., $A(\tau)$) at values of τ that are logarithmically (or near-logarithmically) distributed on the time-scale; as shown in Fig. 9.6(right). In this particular case (taken from (Tassieri et al., 2012)), $A(\tau)$ has been evaluated (or sampled) at lag-times $\tau_n = \text{ceil}(1.45^n)$, for non-negative integer n (where ceil $(\ldots)$ is the ceiling function, which rounds the input variable to the next highest integer); so that, the first five points are linearly spaced in

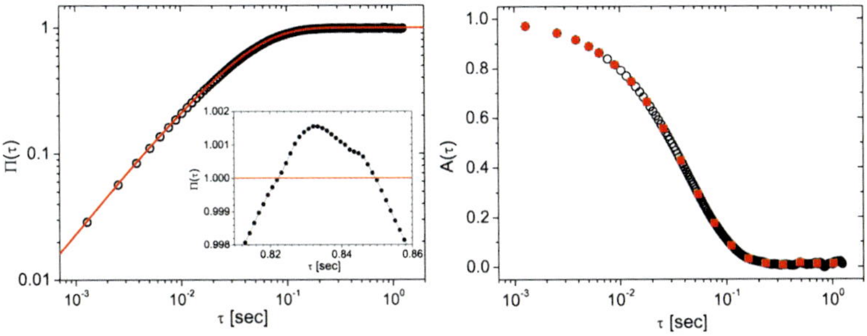

Figure 9.6 (Left) Comparison between the $\Pi(\tau)$ (circles) and its prediction (continuous line), *via* Eqs. 9.24 and 9.21, for an optically trapped 4.74 μm diameter silica bead suspended in water, with $\kappa = 0.93$ μN/m and $\eta = 0.896$ mPa·s. The NMSD has been obtained from the analysis of 10^6 data points representing the particle trajectory, acquired at $AR \simeq 1$ kHz. The inset highlights the behaviour of $\Pi(\tau)$ within a small time-window taken at long lag-times. (Right) The same data as shown on the left, but plotted in terms of the normalised position autocorrelation function *vs.* lag-time (circles); whereas the red square symbols represent the same data as before, but sampled at lag-times quasi-logarithmically distributed on the time-scale: $\tau_n = ceil(1.45^n)$, for non-negative integer *n*.

time, whereas all the others are logarithmically distributed. In this way, the statistical weight of the reliable data at short lag-times is enhanced and the number of disruptive short time-scale deviations of $A(\tau)$ from its expected value, occurring at long lag-times are substantially reduce.

In conclusion, in order to obtain the full viscoelastic spectrum of optically trapped micro-spheres suspended into a generic fluid, both the above data manipulations must by sequentially applied, but in the inverse order as described. The method described above has been successfully applied to both Newtonian and complex fluids, with good agreement when compared to conventional bulk rheology measurements

For example, in the case of Newtonian fluids, Fig. 9.7(left) shows the normalised position autocorrelation functions of optically trapped spheres suspended in both water and 20% w/w glycerol/water mixture; while, in Fig. 9.7(right), are shown the real and the imaginary parts of the normalised complex modulus evaluated *via* Eq. 9.20 and by means of Eq. 9.23 applied to the sampled data

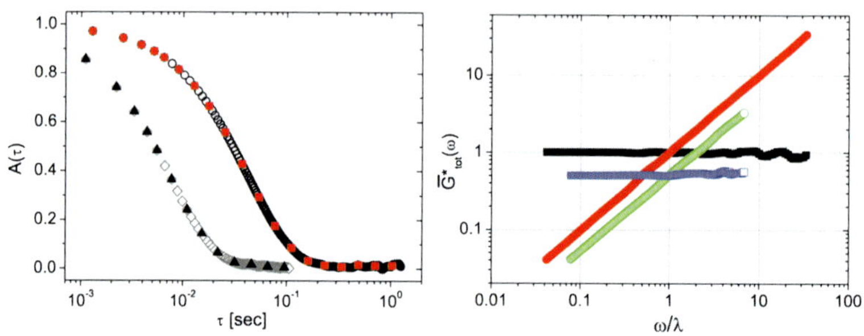

Figure 9.7 (Left) Comparison between the normalised position autocorrelation function shown in Fig. 9.4 (circles) with that obtained from the analysis of an optically trapped 2 μm diameter silica bead suspended in 20% w/w glycerol/water mixture (diamonds), with $\kappa = 4$ μN/m and $\eta = 1.6$ mPa·s. Both the NPAFs have been sampled at lag-times quasi-logarithmically distributed on the time-scale: $\tau_n = ceil(1.45^n)$, for non-negative integer n (squares and triangles, respectivelly). (Right) The normalised complex modulus *vs.* frequency (both dimensionless) evaluated *via* Eq. 9.20 and by means of Eq. 9.23 applied to the sampled NPAFs shown on the left, but interpolated with a natural cubic spline function having $f_s = 25$ MHz and $f_s \simeq 3.7$ MHz, respectively. Note that, the normalised moduli obtained from the glycerol/water mixture (i.e., $\bar{G}'$, blue open square symbols, and $\bar{G}''$, green open circle symbols) have been scaled by a factor of two for a clearer visualisation. The black square and the red circle symbols represent the normalised moduli for the measurement performed with water.

shown in Fig. 9.7(left) after interpolation with a natural cubic spline function having $f_s = 25$ MHz and $f_s \simeq 3.7$ MHz, respectively. From Fig. 9.7(right) it is clear that, although there remains some noise in the real component of the complex modulus at high frequencies, the agreement between the results and the expected values (i.e., Eq. 9.26) is good.

Based on these results, it is possible to confirm that the OT acts as a linear force transducer when operating in the range of frequencies up to ~kHz and on micron-sized particles (i.e., when the laser wavelength is smaller than the particle diameter). However, it is important to be aware that, under other operating conditions, the OT response may not remain linear (Franosch et al., 2011; Jannasch et al., 2011; Pesce et al., 2009; Wu et al., 2009).

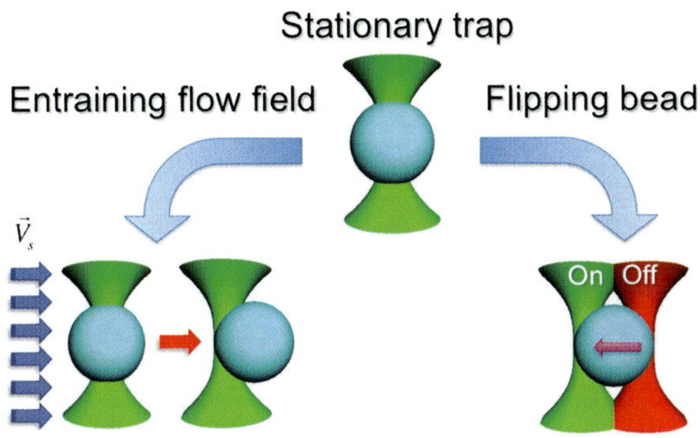

Figure 9.8 Schematic representation of the two experimental procedures introduced by (Tassieri et al., 2010a) and (Preece et al., 2011) for for measuring the viscoelastic properties of complex fluids having the longest relaxation time (τ_D) bigger than a characteristic time of the compound system (OT, bead, fluid) $\lambda^{-1} = (6\pi a\eta_0)/\kappa$, where η_0 is the zero shear viscosity of the fluid.

9.4 Active Microrheology with Optical Tweezers

When optical tweezers are adopted for measuring the viscoelastic properties of complex fluids having the longest relaxation time (τ_D) bigger than a characteristic time of the compound system (OT, bead, fluid) $\lambda^{-1} = (6\pi a\eta_0)/\kappa$, where η_0 is the zero shear viscosity of the fluid, the results are likely to be either limited to the material high frequency response (Atakhorrami et al., 2006; Nijenhuis et al., 2009; Starrs and Bartlett, 2003), discarding the essential information related to long times scales (i.e., low frequency) material behaviour, or supported by low frequency measurements performed by different techniques (e.g., rotational rheometry (Pesce et al., 2009) or passive video particle tracking microrheology (Nijenhuis et al., 2009)), but often either without showing a clear overlapping region between the results (Pesce et al., 2009) or even leaving a macroscopic gap of information in the range of frequencies explored (Nijenhuis et al., 2009).

Two solutions to the above issue have been presented by (Tassieri et al., 2010a) and (Preece et al., 2011). In particular, they presented two self-consistent procedure for measuring the linear viscoelastic properties of materials, from non-oscillatory measurements, across the widest frequency range achievable with optical tweezers, as described hereafter.

9.4.1 *Entraining Flow Field*

The first experimental procedure introduced by (Tassieri et al., 2010a) consists of two steps: (I) measuring the thermal fluctuations of a trapped bead for a sufficiently long time; (II) measuring the transient bead displacement, from the optical trap centre, in response to a uniform fluid flow field entraining the bead. The flow is instantaneously switched on at time zero, by translating the whole fluid sample while the trap is held fixed. The imposed constant-velocity motion continues until a steady displacement of the bead is reached. The analysis of the first step (I) provides: (a) the trap stiffness (κ)—notice that this has the added advantage of making the method self-calibrated—and (b) the high frequency viscoelastic properties of the material as described initially in this chapter. The second step (II) has the potential to provide information about the viscoelastic properties of the material down to very low frequencies, limited only by the duration of the experiment. However, because of the harmonic nature of the optical trap, that tends not to trasmit high-frequency applied forces to the bead, the material's high-frequency response can not be determined by this step. The full material viscoelastic spectrum is thus resolved by combining the results obtained from steps (I) and (II).

The experimental procedure is analytically described through the analysis of the motion of a bead trapped in a stationary harmonic potential of force-constant κ, where a uniform fluid flow field of magnitude $\left|\vec{V}_s\right|$ can be exerted at time $t = 0$. The equation describing the bead position $\vec{r}(t)\ \forall\ t$ can be derived by means of a generalised Langevin equation very similar to that written for the static case (i.e., Eq. 9.5), but with an extra term, $\vec{V}_s(t)$, that takes into a count the velocity of the fluid flow field:

$$m\vec{a}(t) = \vec{f}_R(t) - \int_0^t \zeta(t-\tau)(\vec{v}(\tau) - \vec{V}_s(\tau))d\tau - \kappa\vec{r}(t) \qquad (9.27)$$

In order to derive the fluid's LVE properties, Eq. 9.27 can be solved independently for both the cases mentioned above: i.e., when $\vec{V}_s(t) = \vec{0}$ and $\vec{V}_s(t) \neq \vec{0}$, respectively. In particular, for the first case, Eq. 9.27 returns Eq. 9.5 for which the solution has been already discussed both in Chapter 7 and in Section 9.3.1; whereas, the second case is described as it follows.

The second step (i.e., $\vec{V}_s(t) \neq \vec{0}$), which experimentally follows the first, consists of the analysis of the induced bead displacement from the trap centre due to an imposed time-dependent uniform fluid flow field $\vec{V}_s(t)$ entraining the bead. Equation 9.27 can be solved in terms of the particle position $\vec{r}(t)$ and for simplicity of notation it is rewritten without the vectorial symbol on the variables:

$$ma(t) = f_R(t) - \int_0^t \zeta(t-\tau)(v(\tau) - V_s(\tau))d\tau - \kappa r(t) \qquad (9.28)$$

The first mathematical step is to perform the Laplace transform of Eq. 9.28:

$$\begin{aligned} m\left[s^2\tilde{r}(s) - sr(t_0) - \dot{r}(t_0)\right] &= \tilde{f}_R(s) \\ &- \tilde{\zeta}(s)\left[s\tilde{r}(s) - r(t_0) - \tilde{V}_s(s)\right] - \kappa\tilde{r}(s) \end{aligned} \qquad (9.29)$$

where $\dot{r}(t_0)$ is the time derivative of the particle position evaluated at the instant at which the flow starts, $t = t_0$: i.e., $\dot{r}(t_0) = \left.\frac{dr(t)}{dt}\right|_{t_0}$.

Let us take the average over several independent measurements (but not averaged over absolute time, since time-translation invariance has been broken by the flow start-up at $t = t_0$):

$$\begin{aligned} m\left[s^2\langle\tilde{r}(s)\rangle - s\langle r(t_0)\rangle - \langle\dot{r}(t_0)\rangle\right] &= \langle\tilde{f}_R(s)\rangle \\ &- \tilde{\zeta}(s)\left[s\langle\tilde{r}(s)\rangle - \langle r(t_0)\rangle - \langle\tilde{V}_s(s)\rangle\right] - \kappa\langle\tilde{r}(s)\rangle \end{aligned} \qquad (9.30)$$

where the averaged terms $\langle r(t_0)\rangle$, $\langle\dot{r}(t_0)\rangle$ and $\langle\tilde{f}_R(s)\rangle$ are all equal to zero. This is because, over all the repeated measurements, the first of them is the average particle position before the flow begins, which turns to be coincident with the origin of the coordinate system ($\langle\vec{r}(t_0)\rangle \equiv \vec{0}$); whereas, the latter two are variables that can arbitrarily assume any value at time $t = t_0$, thus their averages are equal to zero. Therefore, Eq. 9.30 becomes:

$$ms^2\langle\tilde{r}(s)\rangle = -\tilde{\zeta}(s)\left(s\langle\tilde{r}(s)\rangle - \langle\tilde{V}_s(s)\rangle\right) - \kappa\langle\tilde{r}(s)\rangle \qquad (9.31)$$

from which it is an easy step to show that:

$$\tilde{\xi}(s) = \frac{(ms^2 + \kappa)\langle\tilde{r}(s)\rangle}{(\langle\tilde{V}_s(s)\rangle - s\langle\tilde{r}(s)\rangle)} \tag{9.32}$$

Then, based on Eqs. 9.6 and 9.17, it is straightforward to show that the fluid's complex modulus can be expressed directly in terms of both the imposed flow field and of the induced bead displacement from the trap centre:

$$G^*(\omega) = \frac{(\kappa - m\omega^2)i\omega\langle\hat{r}(\omega)\rangle}{6\pi a\left(\hat{V}_s(\omega) - i\omega\langle\hat{r}(\omega)\rangle\right)} \tag{9.33}$$

where $\hat{V}_s(\omega)$ and $\langle\hat{r}(\omega)\rangle$ are the Fourier transforms of $\vec{V}_s(t)$ and $\langle\vec{r}(t)\rangle$, respectively. Moreover, the averaged term $\langle\hat{V}_s(\omega)\rangle$ has been replaced with $\hat{V}_s(\omega)$, because it has been assumed that within a set of repeated measurements $\vec{V}_s(t)$ would not change. Also in this case, for micron sized particles the term $m\omega^2$ is negligible compared to κ for frequencies $\omega <<$ MHz; and Eq. 9.33 can be reorganised as:

$$G^*(\omega)\frac{6\pi a}{\kappa} = \left(\frac{\hat{V}_s(\omega)}{i\omega\langle\hat{r}(\omega)\rangle} - 1\right)^{-1} \equiv \left(\frac{\hat{V}_s(\omega)}{\langle\hat{v}(\omega)\rangle} - 1\right)^{-1} \tag{9.34}$$

Where $\hat{v}(\omega) = i\omega\hat{r}(\omega)$ is the Fourier transform of the particle velocity $v(t) = dr/dt$, thus the substitution in Eq. 9.34. It is important to note that, so far, the temporal form of $\vec{V}_s(t)$ is still undefined. Thus, Eq. 9.33 represents the general solution for $G^*(\omega)$ independently of the temporal form of $\vec{V}_s(t)$ (e.g. sinusoidal function $\vec{V}_s \sin(\omega t)$ or, as in (Tassieri et al., 2010a), Heaviside step function $\vec{V}_s H(t)$, where $\vec{V}_s(t) = \vec{0}\ \forall\ t < 0$ and $\vec{V}_s(t) = \vec{V}_s\ \forall\ t \geqslant 0$).

The above experimental procedure (i.e., the ensemble of steps I and II) has been validated in (Tassieri et al., 2010a) by measuring both the viscosity of water and the viscoelastic properties of water-based solutions of polyacrylamide, a flexible polyelectrolytes (PAM, $M_w = 5\text{–}6 \times 10^6$ g/mol, Polysciences Inc.) as described below.

The Brownian fluctuations of an optically trapped bead give rise to the time dependent $\langle\Delta r^2(\tau)\rangle$ shown in Fig. 9.9. In the case of a bead immersed in a Newtonian fluid, it is expected that at short time intervals (thus small distances) the bead behaves as if it were free to diffuse. Indeed, the agreement shown in Fig. 9.9 between the observed $\langle\Delta r^2(\tau)\rangle$ at short times of a trapped bead in water (circles)

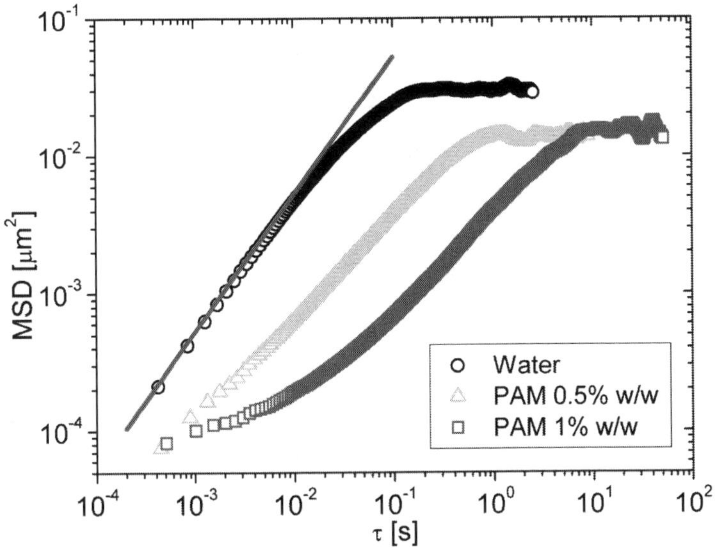

Figure 9.9 The MSD *vs.* lag-time of a 5 μm diameter bead in water (with $\kappa = 0.8$ μN/m) and in two water based solutions of PAM at concentrations of 0.5% w/w and 1% w/w (both with $\kappa = 1.7$ μN/m). The line is the Einstein prediction of the MSD for a freely diffusing 5 μm diameter bead in water at 25°C. Image reproduced from (Tassieri et al., 2010a).

and the Einstein prediction for a freely diffusing bead (solid line) is good. As the time intervals increase the bead becomes influenced by the optical potential. This results in a plateau at large time intervals, where the $\langle \Delta r^2(\tau) \rangle$ tends to $2\langle r^2 \rangle_{eq}$; thus the definition of the normalised mean squared displacement $\Pi(\tau)$, i.e., Eq. 9.7 (see Fig. 9.10). It is interesting to note that the onset point of the plateau region shown in Fig. 9.10 indicates (by means of the inverse of its abscissa) the bottom limit of the frequency range within which the moduli can be determined by Eq. 9.20; as for all methods using stationary optical tweezers.

In Fig. 9.11 are compare the responses of a 5 μm diameter bead immersed in water (a Newtonian fluid) and in a water solution of PAM at 1% w/w (a non-Newtonian fluid), due to the imposition of a uniform fluid flow field having temporal behaviour as a Heaviside step function $\vec{V}_s(t) = \vec{V}_s H(t)$, with different magnitude in the two measurements. Experimentally, the execution of a Heaviside step

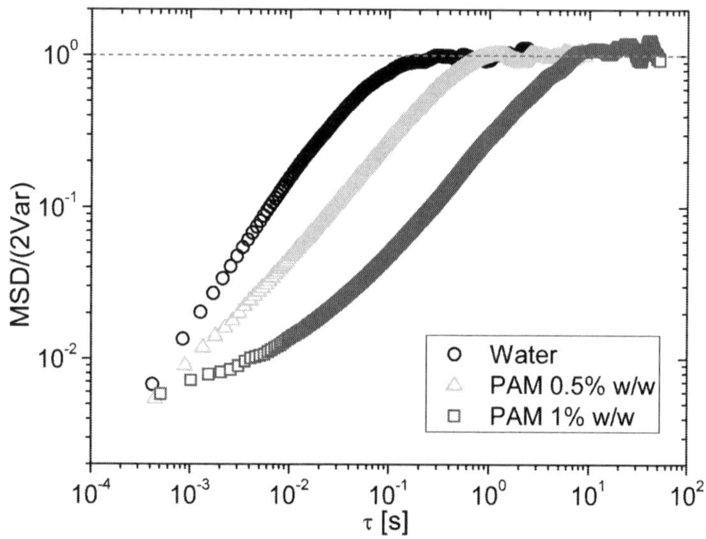

Figure 9.10 The normalised MSD *vs.* lag-time of a 5 μm diameter bead in water (with $\kappa = 0.8$ μN/m) and in two water based solutions of PAM at concentrations of 0.5% w/w and 1% w/w (both with $\kappa = 1.7$ μN/m). Image reproduced from (Tassieri et al., 2010a).

function is achieved by suddenly moving the motorised microscope stage at a predetermined speed and direction (here parallel to the x axis). The experiment runs until a steady displacement (Δx) of the bead from the trap centre is reached (i.e., until all the material's characteristic relaxation times are exceeded). In Fig. 9.11 the x component of the bead displacement has been normalised by Δx for a better comparison between the viscoelastic character of the two samples. It is clear that while the Newtonian fluid reaches a steady value of the displacement almost *instantaneously* (as expected), the non-Newtonian fluid shows complex dynamics representative of its viscoelastic nature. It is important to note that, because of the harmonic nature of the optical potential, at early times (i.e., for $t \to 0$ or equivalently for $\omega \to \infty$), the trapping force exerted on the bead is actually small (i.e., $\kappa \vec{r}(t) \to 0$) and the particle moves almost at the same speed as the imposed flow (i.e., $\vec{v}(t) \cong \vec{V}_s(t)$); this implies that Eq. 9.34 becomes undefined at high frequencies.

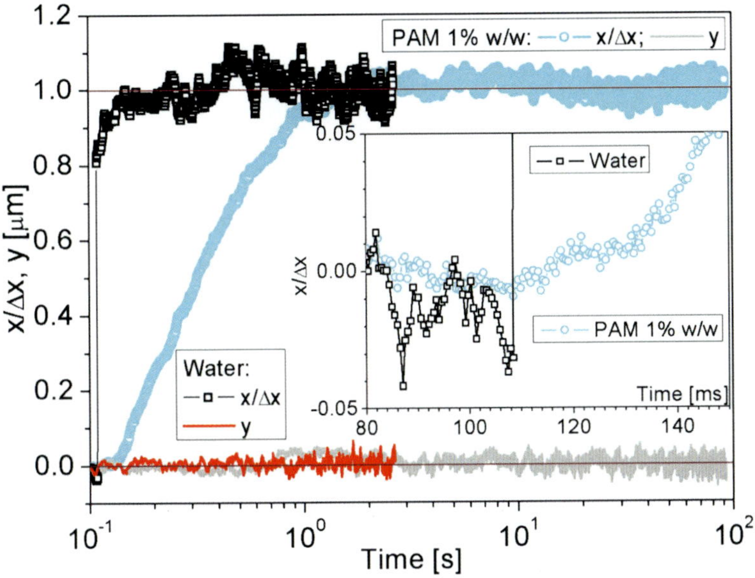

Figure 9.11 The coordinates of a 5 μm diameter bead *vs.* time for two different solutions and for two uniform fluid flow fields of different magnitude $|\vec{V}_s|$ at 25°C. In both the cases the data were averaged over three measurements and the x coordinate has been normalised by the steady state displacement Δx. In water: $\kappa = 1.7$ μN/m, $|\vec{V}_s| = 20$ μm/s and $\Delta x = 0.523$ μm. In 1% w/w of PAM: $\kappa = 8.6$ μN/m, $|\vec{V}_s| = 3$ μm/s and $\Delta x = 1.155$ μm. The inset highlights the start-up behaviour of both the above systems. Image reproduced from (Tassieri et al., 2010a).

The broadband microrheology measurement with optical tweezers is achieved by combining the frequency responses obtained from both the methods introduced above (i.e., step I *plus* step II). In particular, the material's high frequency response is determined by applying Eq. 9.20 (*via* Eq. 9.23 with $\langle \Delta r^2 \rangle_k$ replacing g_k) to the $\langle \Delta r^2(\tau) \rangle$ measurements; whereas, the low frequency response is resolved by applying Eq. 9.34 (*via* Eq. 9.23 with $\langle \vec{r} \rangle_k$ replacing g_k) to the data describing the bead's transient response to the motion of the stage.

A typical result of this procedure for a non-Newtonian fluid is shown in Fig. 9.12. It is evident that, although there is some noise in the frequency domain that has propagated from genuine

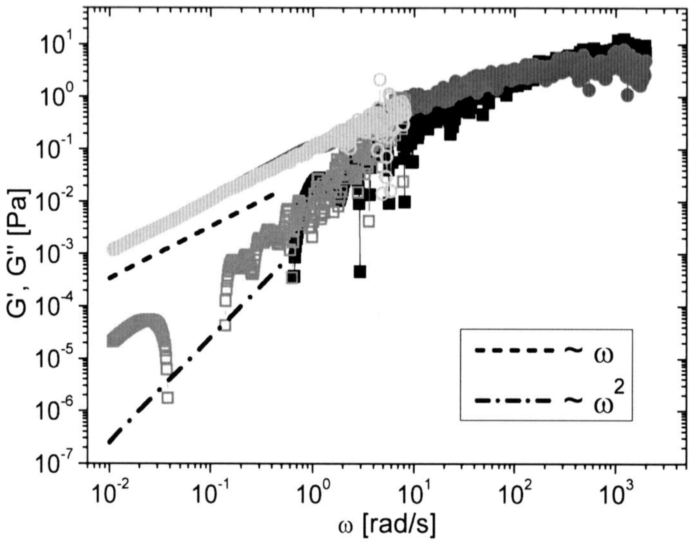

Figure 9.12 Storage (squares) and loss (circles) moduli *vs.* frequency of a solution of 1% w/w of PAM in water measured by means of both Eq. 9.20 (solid symbols at high frequencies) and Eq. 9.34 (open symbols at low frequencies) applied directly to the experimental data presented in Figs. 9.10 and 9.11, respectively. Image reproduced from (Tassieri et al., 2010a).

experimental noise in the time-domain data (because in this particular case the data analysis was missing the analytical procedure described in Section 9.3.2, which was introduced by (Tassieri et al., 2012) in a later stage), there is a clear overlapping region of agreement between the two methods that makes the whole procedure self-consistent. Moreover, it confirms the ease with which the low-frequency material response can be explored, right down to the terminal region (where $G' \propto \omega^2$ and $G'' \propto \omega$). The experimental method presented by (Tassieri et al., 2010a) has no fundamental restriction on the lowest achievable frequency. Indeed, with a suitable choice of the velocity of the applied fluid flow field (e.g., $\left|\vec{V}_s\right| = 3$ μm/s) and the sample holder geometry (e.g., a 3 cm long microfluidic channel), an experiment could be made to run for hours, thereby probing frequencies of order 10^{-4} s^{-1} or lower, without limit.

9.4.2 *Flipping Bead*

An alternative experimental procedure to that introduced by (Tassieri et al., 2010a) for measuring the linear viscoelastic properties of fluids having $\tau_D > \lambda^{-1}$ is the one introduced by (Preece et al., 2011). They also presented a self-consistent procedure for measuring the LVE properties of materials, from non-oscillatory measurements, across the widest frequency range achievable with optical tweezers. In particular, the proposed procedure consists of two steps: (I) measuring the thermal fluctuations of a trapped bead for a sufficiently long time; (II) measuring the transient displacement of a bead flipping between two optical traps (spaced at fixed distance D_0) that alternately switch on/off at sufficiently low frequency. The analysis of the first step (I) provides: (a) the traps stiffness (κ_i, $i = 1, 2$)—notice that this has the added advantage of making the method self-calibrated—and (b) the high frequency viscoelastic properties of the material, to high accuracy. The second step (II) has the potential to provide information about the material's viscoelastic properties over a very wide frequency range, which is only limited (at the top end) by the acquisition rate of the bead position ($\sim$kH) and (at the bottom end) by the duration of the experiment. However, because of the finite time required by the equipment to switch on/off (i.e., tens of milliseconds), the material's high-frequency response can not be fully determined by this step. The full viscoelastic spectrum is thus resolved by combining the results obtained from steps (I) and (II).

Also in this case, the experimental procedure is analytically described through the analysis of the motion of a bead trapped by a stationary harmonic potential of force-constant κ. The equation describing the bead position $\vec{r}(t)$ $\forall$ t can be derived by means of a generalised Langevin equation identical to Eq. 9.5. In order to derive the fluid's LVE properties, Eq. 9.5 can be solved independently for both the cases mentioned above. In particular, for the first case, the solution has been already discussed both in Chapter 7 and in Section 9.3.1; whereas, the second case is described as it follows.

Throughout the first step (I) of the procedure, the bead is always found close to the centre of the single trap which is switched on; it makes only small deviations $\vec{r}(t)$ (of magnitude set by the thermal energy) away from the centre of the trap. During step (II) of the procedure, the detours are considerably larger, of a magnitude set by the separation D_0 of the traps. Let us define t_{II} to be the time at which step (II) commences, i.e., the time at which the traps are first switched. Subsequently, each trap remains on for a duration P before it is switched off and the other trap switched on. Hence the total period of the repeated sequence is $2P$. At the instant immediately after the traps are switched (i.e., at $t = t_{\text{II}} + nP$ where $n = 0, 1 \ldots N$), the bead is typically positioned at a distance $|\vec{r}(t)| \simeq D_0$ from the centre of the currently active trap (i.e., close to the centre of the trap that has just switched off). It is important to note that coordinates are re-defined so that the bead's displacement $\vec{r}(t)$ is always measured with respect to the centre of whichever trap is currently switched on. It follows that:

$$D_0 \equiv \langle \vec{r}(t_{\text{II}} + nP) \rangle_N, \tag{9.35}$$

which is the mean position of the particle at times $t = t_{\text{II}} + nP$ (i.e., when the traps swap their on/off state), when the average is taken over all the (N) repeated sequences of duration P.

The analytical procedure consists of analysing the bead's transient displacements as it moves between two traps with separation D_0 that swap their on/off state at times $t = t_{\text{II}} + nP$. It is important to note that the duration P must be much longer than τ_{D}. It is possible to define a normalised mean position of the particle as:

$$D(t) = \frac{\langle \vec{r}(t) \rangle}{D_0} \tag{9.36}$$

where the brackets $\langle \ldots \rangle$ denote the average over several independent measurements, but not over absolute time, since time-translation invariance is broken by the periodic switching.

Let us show step by step how Eq. 9.5 can be solved in terms of the $D(t)$ and for simplicity let us rewrite it without the vectorial notation:

$$m\ddot{r}(t) = f_R(t) - \int_0^t \zeta(t-\tau)\dot{r}(\tau)d\tau - \kappa r(t) \tag{9.37}$$

where $\dot{r}(t)$ and $\ddot{r}(t)$ denote respectively the first and the second time derivative of the particle position $r(t)$. The first step is to perform the Laplace transform of Eq. 9.37:

$$m[s^2\tilde{r}(s) - sr(t_{\mathrm{II}}) - \dot{r}(t_{\mathrm{II}})] = \tilde{f}_R(s) - \tilde{\xi}(s)[s\tilde{r}(s) - r(t_{\mathrm{II}})] - \kappa\tilde{r}(s) \quad (9.38)$$

where s is the Laplace frequency and the convolution integral is transformed into the product of the two Laplace-transformed integrand functions. The following step is to take the average over all the repeated sequences:

$$m[s^2\langle\tilde{r}(s)\rangle - s\langle r(t_{\mathrm{II}})\rangle - \langle\dot{r}(t_{\mathrm{II}})\rangle] = \left\langle\tilde{f}_R(s)\right\rangle - \tilde{\xi}(s)[s\langle\tilde{r}(s)\rangle - \langle r(t_{\mathrm{II}})\rangle] - \kappa\langle\tilde{r}(s)\rangle \quad (9.39)$$

where the averaged terms $\langle\dot{r}(t_{\mathrm{II}})\rangle$ and $\left\langle\tilde{f}_R(s)\right\rangle$ are equal to zero. This is because, over all the repeated measurements (*or* sequences), these variables can arbitrarily assume any value at time $t = t_{\mathrm{II}}$, thus their averages are equal to zero. Therefore, from Eqs. 9.39, 9.35 and 9.36 we can write:

$$ms\left(s\tilde{D}(s) - 1\right) = -\tilde{\xi}(s)\left(s\tilde{D}(s) - 1\right) - \kappa\tilde{D}(s) \quad (9.40)$$

where $\tilde{D}(s)$ is the Laplace transform of the particle normalised mean position $D(t)$. Moreover:

$$ms = -\tilde{\xi}(s) - \kappa\frac{\tilde{D}(s)}{s\tilde{D}(s) - 1} \quad (9.41)$$

from which

$$\tilde{\xi}(s) = \kappa\frac{\tilde{D}(s)}{1 - s\tilde{D}(s)} - ms. \quad (9.42)$$

By combining Eqs. 9.6 and 9.42 we can write:

$$\tilde{\eta}(s) = \frac{\kappa}{6\pi a}\left(\frac{\tilde{D}(s)}{1 - s\tilde{D}(s)} - \frac{ms}{\kappa}\right) \quad (9.43)$$

which is an identical expression to Eq. 9.16 with $\tilde{A}(s)$ replaced by $\tilde{D}(s)$. Finally, from Eqs. 9.43 and 9.17 the complex modulus $G^*(\omega)$ can be expressed directly in terms of the time-dependent normalised mean position of the particle $D(t)$:

$$G^*(\omega) = \frac{\kappa}{6\pi a}\left(\frac{i\omega\hat{D}(\omega)}{1 - i\omega\hat{D}(\omega)} + \frac{m\omega^2}{\kappa}\right) \quad (9.44)$$

where $\hat{D}(\omega)$ is the Fourier transform of $D(t)$.

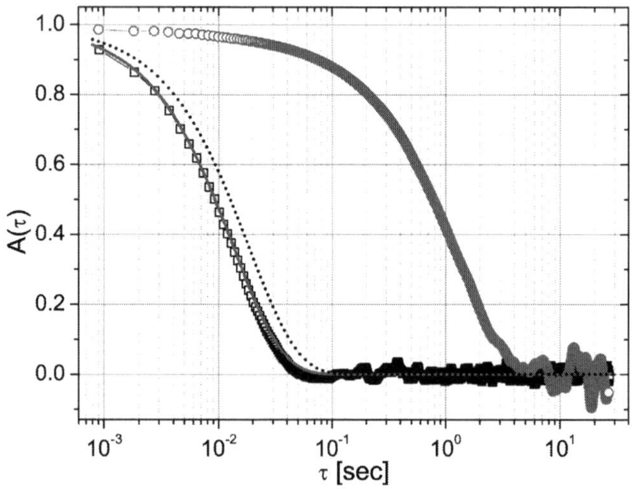

Figure 9.13 The normalised position autocorrelation function *vs.* lag-time of a 5 μm diameter bead (squares) in water (with $\kappa = 2.7$ μN/m) and (circles) in a water-based solution of PAM at concentrations of 1% w/w (with $\kappa = 2.2$ μN/m). The continuous and dotted lines represent Eq. 9.24 for a 5 μm diameter bead in water at $T = 25°$C with $\kappa = 2.7$ μN/m and $\kappa = 2.2$ μN/m, respectively. Image reproduced from (Preece et al., 2011).

In the case where the two traps had different stiffnesses κ_j (with $j = 1, 2$), Eq. 9.44 can be reorganised as:

$$G^*(\omega) = \frac{1}{12\pi a} \sum_{j=1,2} \left[\frac{\kappa_j i\omega \hat{D}(\omega)}{\left(1 - i\omega \hat{D}(\omega)\right)} + m\omega^2 \right], \tag{9.45}$$

where the sum takes account of the linearity of the measurements performed with both traps.

It is important to note that both functions $A(\tau)$ and $D(t)$ are expected to have the limits $A(0) = D(0) = 1$ and $A(\infty) = D(\infty) = 0$. This turns out to be very useful when applying the Fourier transform during data processing, because of the procedure described in Appendix.

The above experimental procedure (i.e., the ensemble of steps I and II) has been validated in (Preece et al., 2011) by measuring both the viscosity of water and the viscoelastic properties of water-based solutions of polyacrylamide (PAM, $M_w = 5–6 \times 10^6$ g/mol, Polysciences Inc.) and their results are described below.

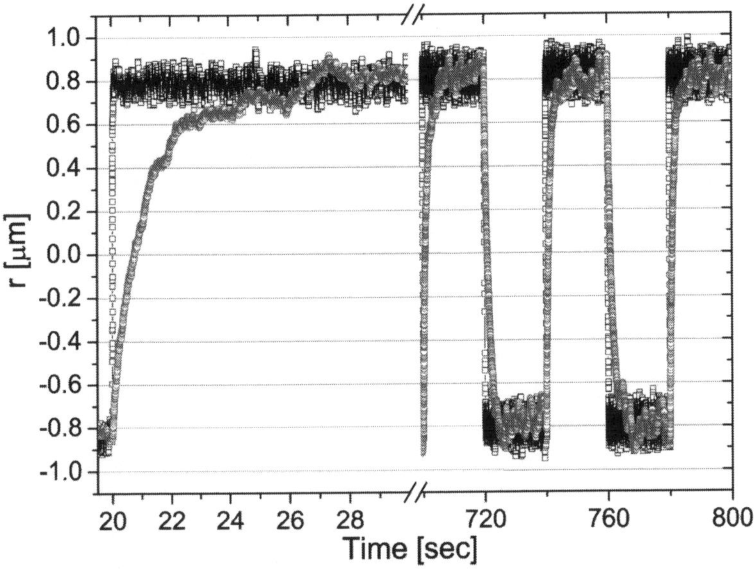

Figure 9.14 The trajectory of a 5 μm diameter bead flipping between two optical traps κ_1 (bottom) and κ_2 (top) repeatedly switching after a duration $P = 20$ s. The bead is suspended in (squares) water (with $\kappa_1 = 2.7$ and $\kappa_2 = 2.5$ μN/m) and (circles) a water-based solution of PAM at concentrations of 1% w/w (with $\kappa_1 = 2.1$ and $\kappa_2 = 2.2$ μN/m). Image reproduced from (Preece et al., 2011).

Figure 9.13 shows the normalised position autocorrelation functions $A(\tau)$, measured from the stochastic fluctuations of beads optically trapped in both water and a water-based solution of PAM at concentrations of 1% w/w. In the same figure, the theoretical predictions (i.e., Eq. 9.24) are drawn as lines for two systems having the same Newtonian viscosity of water, but trap stiffnesses like those of the above experimental measurements. In the case of water, the agreement between the experimental data and the prediction is good. Whereas, in the case of the non-Newtonian fluid, the mismatch between the theoretical (dotted) curve and the experimental data is indicative of the viscoelastic contribution provided by the solute to the solvent. The LVE properties of the solution can be determined by means of Eq. 9.20. In Fig. 9.14 are compared the impulse response (i.e., step II of the procedure) of a 5 μm diameter bead suspended in water and in an aqueous solution of PAM at 1% w/w, with a duration between flips of $P = 20\ s$ and a trap centre-to-centre separation of $D_0 = 1.6\ \mu\text{m}$, giving $D_0/a = 0.64$. In order to guarantee the linearity of the confining forces exerted by the two optical traps, the distance between them was always chosen to be no more than 80% of the bead radius, $D_0 \leqslant 0.8a$ (Ashkin, 1992). Although Brownian fluctuations appear in the figure, the difference between the viscoelastic natures of the two fluids is clear. Indeed, while a bead suspended in water flips from one trap to the other almost instantaneously, because of the fluid's low viscosity, the same bead in the PAM solution takes much longer (a few seconds) to flip (because of the fluid's relatively higher viscoelasticity). In order to reduce noise caused by the Brownian fluctuations and thus evaluate $D(t)$, the transient measurements were averaged over twenty flips, with the resulting curves shown in Fig. 9.15. Experimentally, the switching process of the two traps is controlled by means of a spatial light modulator (SLM) which alternately creates an optical trap in

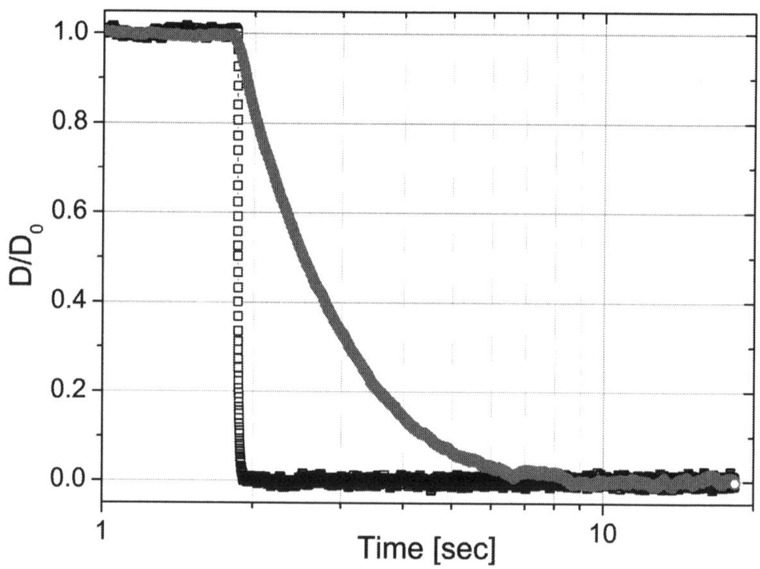

Figure 9.15 The normalised mean position of all step-down data shown in Fig. 9.14; i.e., when simultaneously trap 2 (top) switches *off* and trap 1 (bottom) switches *on*. Image reproduced from (Preece et al., 2011).

one of two positions. It is important at this point to note that there is a short but finite time for which both traps exist simultaneously (Preece et al., 2009), due to the finite time required by the SLM's display to update the holographic pattern. This makes the switching process not exactly binary. However, this will of course only affect the high-frequency results obtained during this step which will ultimately be neglected in favour of the high-frequency response from step (I).

Wideband microrheology measurements with optical tweezers are achieved by combining the frequency responses obtained from both steps (I) and (II) of the procedure. In particular, the material's high-frequency response is determined by applying Eq. 9.20 (*via* Eq. 9.23 with A_k replacing g_k) to the $A(\tau)$ measurements (in which low-frequency information are overshadowed by the trap stiffness); whereas, the low-frequency response is resolved by applying Eq. 9.44 (*via* Eq. 9.23 with D_k replacing g_k) to the data describing the bead's transient response to the switching traps (in which the high-frequency response is limited by the performance of the SLM).

Typical results for both Newtonian and non-Newtonian fluids are shown in Figs. 9.16 and 9.17, respectively. In both the cases, it is evident that, although there is some noise in the frequency domain, which comes from genuine experimental noise in the time-domain data (because the data analysis was missing the analytical procedure described in Section 9.3.2), there is a clear overlapping region of agreement between the two steps. This confirms that the whole procedure is self-consistent, and allays any possible concerns about the linearity or reliability of low frequency measurements (Schnurr et al., 1997). Moreover, it confirms the ease with which the low-frequency material response can be explored right down to the terminal region (where $G' \propto \omega^2$ and $G'' \propto \omega$).

9.5 A Rheological Interpretation of Optical Tweezers

It is worthwhile to review some fundamental relationships between the most common parameters describing the materials' LVE

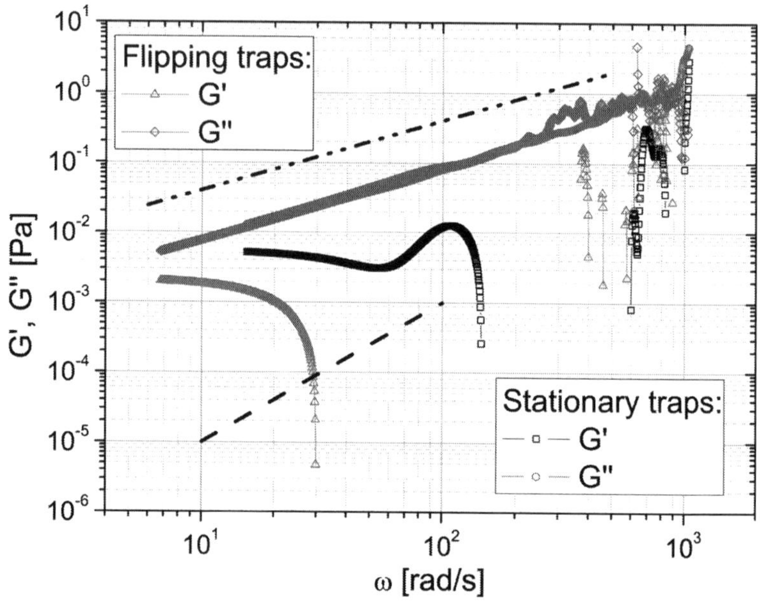

Figure 9.16 Storage (G') and loss (G'') moduli of water *vs.* frequency, analysed using both Eq. 9.20 (high frequencies) and Eq. 9.44 (low frequencies) applied directly to the experimental data presented in Figs. 9.13 and 9.15, respectively. The lines represents the expected limiting behaviour of the moduli when the material reaches the terminal region: $G' \propto \omega^2$ and $G'' \propto \omega$. Image reproduced from (Preece et al., 2011).

properties and the time-averaged functions (e.g. MSD) derived by the analysis of the particle's thermal fluctuations.

Let us begin from the simple relationship between the MSD of a freely diffusing particle and the time-dependent shear creep compliance $J(t)$ of the suspending fluid. In this case, based on the pioneering work of (Mason and Weitz, 1995), we acknowledge the relationship between the Fourier transform of the MSD and the fluid's complex shear modulus:

$$\left\langle \Delta \widehat{r^2}(\omega) \right\rangle = \frac{k_B T}{\pi a i \omega G^*(\omega)} \tag{9.46}$$

where it has been assumed that the inertial term $(m\omega^2)$, which was present in the original derivation of Eq. 9.46, is negligible for frequencies $\omega \ll$ MHz. Moreover, as introduced in Chapter 6

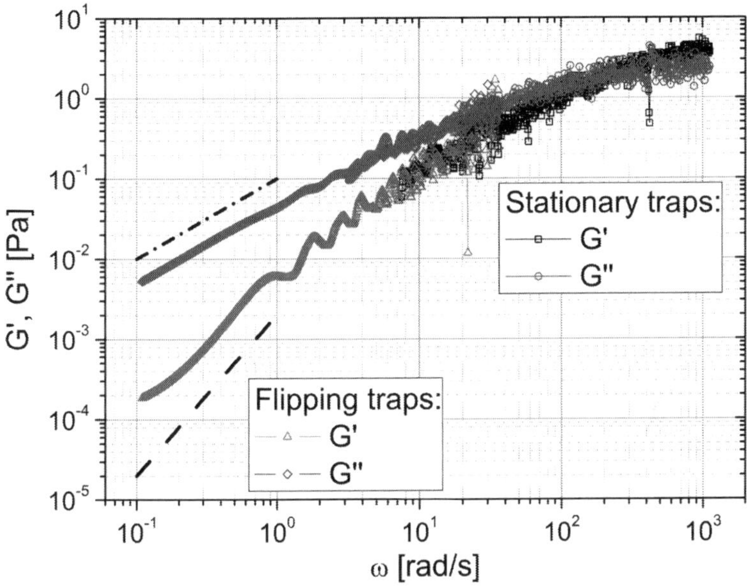

Figure 9.17 Storage (G') and loss (G'') moduli *vs.* frequency of a solution of 1% w/w of PAM in water measured by means of both Eqs. 9.20 (high frequencies) and 9.44 (low frequencies) applied directly to the experimental data presented in Figs. 9.13 and 9.15, respectively. The lines represents the expected limiting behaviour of the moduli when the material reaches the terminal region: $G' \propto \omega^2$ and $G'' \propto \omega$. Image reproduced from (Preece et al., 2011).

and masterly explained in (Ferry, 1980), we remind that the creep compliance is related to the shear relaxation modulus by means of Eq. 6.6, which is here reported for simplicity:

$$\int_0^t G(\tau) J(t-\tau)\, d\tau = t, \tag{9.47}$$

By performing the Fourier transform of Eq. 9.47 one obtains Eq. 6.12:

$$G^*(\omega) = i\omega\hat{G}(\omega) = \frac{1}{i\omega\hat{J}(\omega)} \tag{9.48}$$

where $\hat{G}(\omega)$ and $\hat{J}(\omega)$ are the Fourier transforms of $G(t)$ and $J(t)$, respectively. Therefore, by equating Eqs. 9.46 and 9.48 one obtains:

$$\left\langle \Delta\widehat{r^2}(\omega) \right\rangle = \frac{k_B T}{\pi a} \hat{J}(\omega) \tag{9.49}$$

And because the Fourier transform is a linear operator, the same equality is valid also in the time domain:

$$\left\langle \Delta r^2(\tau) \right\rangle = \frac{k_B T}{\pi a} J(t) \tag{9.50}$$

The above equation expresses the linear relationship between the MSD of suspended spherical particles and the macroscopic shear creep compliance of the suspending fluid (Xu et al., 1998).

Let us now consider an optically trapped spherical particle suspended in a viscoelastic fluid. For such a system, Eq. 9.50 would still hold, but it would describe the relationship between the measured MSD of a constrained particle and the compliance (J_{tot}) of the compound system made up by the optical trap and the viscoelastic fluid:

$$\left\langle \Delta r^2(\tau) \right\rangle = \frac{k_B T}{\pi a} J_{tot}(t), \tag{9.51}$$

which, with reference to Eq. 9.3, can be written as:

$$\Pi(\tau) = \frac{\left\langle \Delta r^2(\tau) \right\rangle}{2\left\langle r^2 \right\rangle_{eq}} = \frac{\kappa}{6\pi a} J_{tot}(t). \tag{9.52}$$

The above equation denotes the linear relationship between the NMSD of an optically trapped spherical particle and the combined action, expresses in terms of $J_{tot}(t)$, of both the OT trap and the macroscopic shear creep compliance of the suspending viscoelastic fluid.

In the simplest case where the suspending fluid is Newtonian, with a time-independent viscosity η, the compound system (OT *plus* Newtonian fluid) can be modelled as an *ideal* Kelvin-Voigt material (see Section 6.3), with elastic constant proportional to the trap stiffness, $\kappa/(6\pi a)$, and viscosity equal to η. In this case, J_{tot} assumes a simple analytical form:

$$J_{tot} = \frac{6\pi a}{\kappa}\left(1 - e^{-\lambda t}\right) \quad \Rightarrow \quad \Pi(\tau) = \left(1 - e^{-\lambda \tau}\right) \tag{9.53}$$

where $\lambda = \kappa/(6\pi a\eta)$ is the relaxation rate of the compound system, known as the "*corner frequency*" when the thermal fluctuations of an optically trapped bead are analysed in terms of the power spectral density (Berg-Sørensen and Flyvbjerg, 2004). From a rheological point of view, λ defines a characteristic time ($t^* = \lambda^{-1}$) at which the time-dependent fluid compliance ($J(t) = t/\eta$) equals the time-independent compliance of the optical trap ($J_{\mathrm{OT}} = 6\pi a/\kappa$)

$$\frac{t^*}{\eta} = \frac{6\pi a}{\kappa} \Rightarrow t^*\lambda = 1 \tag{9.54}$$

as schematically shown in Fig. 9.18 for both Newtonian and viscoelastic fluids, where the abscissa of the intercept between $J(t) = t/\eta(t)$ and J_{OT} provides a reading of t^*.

At this point it is important to remind that, in general viscoelastic fluids may have several characteristic relaxation times because of their complexity. However, for these fluids it is always possible to identify the *longest* relaxation time, commonly named τ_{D}, such that for either sufficiently slow flows (i.e., $\dot{\gamma} \ll \tau_{\mathrm{D}}^{-1}$) *or* low frequency oscillatory measurements (i.e., $\omega \ll \tau_{\mathrm{D}}^{-1}$) the fluids behave as if it were Newtonian with a time-independent viscosity η_0. The latter, indeed, is also known as either the fluid's zero-shear viscosity *or* the fluid's zero-frequency viscosity, respectively. In Fig. 9.18 this translates in a compliance curve that approaches a linear behaviour (i.e., in a log – log plot, the slope of $J(t)$ tends to 1) for $t \gg \tau_{\mathrm{D}}$, where the asymptotic curve is t/η_0.

It is now possible to explain why microrheology measurements performed with static optical tweezers may not be able to provide the full frequency spectrum of the fluid's viscoelastic properties; hence the need of performing active microrheology by adopting either of the two experimental procedures introduced by (Tassieri et al., 2010a) and (Preece et al., 2011). In particular, with regards to Fig. 9.18, in the case of non-Newtonian fluids, two sceneries are possible:

(i) J_{OT} intercepts $J(t)$ at $t^* \gg \tau_{\mathrm{D}}$;
(ii) J_{OT} intercepts $J(t)$ at $t^* \lesssim \tau_{\mathrm{D}}$.

In the first case (i), which is schematically represented in Fig. 9.18 by the green dash-dot curve (for which $t_3^* \gg \tau_{D1}$), the fluid compliance is well approximated by its asymptotic behaviour of $J(t) \cong t/\eta_0$ well before intercepting J_{OT}. This means that, the elastic component of fluid ($G'(\omega)$) is much smaller than the time-independent elastic component of the OT ($\kappa/6\pi a$), which will inevitably dominate the low frequency (i.e., $\forall\ \omega < \lambda \equiv \kappa/6\pi a\eta_0$) elastic response of the

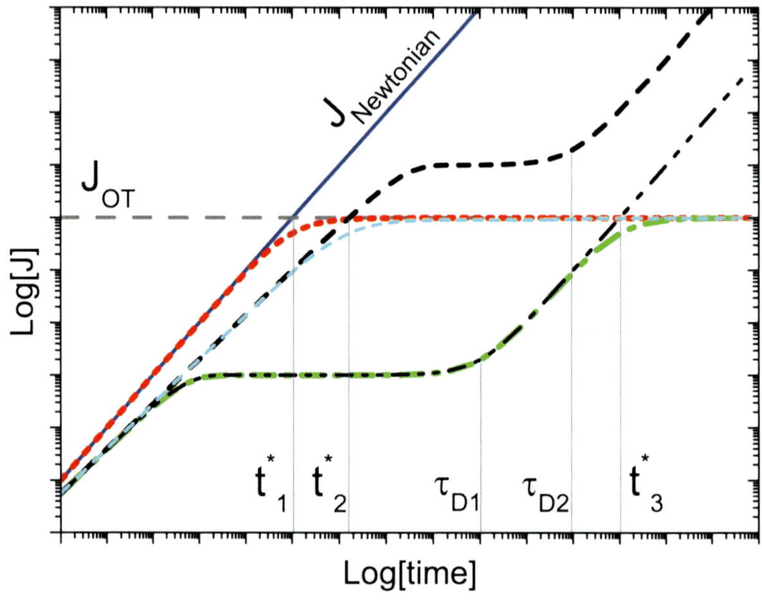

Figure 9.18 Double logarithmic schematic representation of a particle mean-square-displacement in terms of the creep compliance (see Eq. 9.52) suspended into both Newtonian and non-Newtonian fluids with and without optical trapping.

compound system. However, in this case, the fluid's LVE properties are fully determined down to the terminal region (i.e., down to frequencies lower than $1/\tau_D$, where $G'(\omega) \propto \omega^2$ and $G'(\omega) \propto \omega$) by means of the data analysis procedure introduced in Section 9.3. In particular, the fluid's LVE properties can be measured down to a frequency value of $1/t^* < 1/\tau_D$, and they can be easily extrapolated towards lower frequencies because of the known scaling laws of both the moduli within the terminal region.

In the second case (ii), which is schematically represented in Fig. 9.18 by the cyan dash curve (for which $t_2^* \ll \tau_{D2}$), J_{OT} intercepts the fluid compliance curve at a generic point that is far form the fluid's terminal region, where $J(t) \cong t/\eta_0$. This implies that, while the fluid's high frequency (i.e., $\forall\ \omega > 1/t_2^*$) LVE properties could still be determined by means of the data analysis procedure introduced in Section 9.3; at low frequencies, i.e., $\forall\ \omega < 1/t_2^*$, the fluid's LVE properties are *overshadowed* by the time-independent elastic component of the OT (i.e., $\kappa/6\pi a$), with the consequent loss of information. Therefore, in order to evaluate the full frequency spectrum, i.e., to measure the viscoelastic moduli in a range of frequencies delimited at the top-end by the detector acquisition rate of the bead position and at the bottom-end by $1/\tau_{\mathrm{D}}$ *or* lower, one of the two experimental procedures introduced by (Tassieri et al., 2010a) and (Preece et al., 2011) must be employed. However, in both the above cited methods, a particular attention must be paid to the duration of the active part of the measurements (i.e., step (II) of both the methods introduced in Sections 9.4.1 and 9.4.2, respectively), which must exceed τ_{D} and be repeated several times within the whole experiment to reduce the undesired effects of the thermal noise.

Part IV

Review on Optical Tweezers Applications

Chapter 10

Optical Tweezers Outwith Microrheology

Richard W. Bowman
Department of Physics, University of Cambridge, CB3 0HE, UK
Queens' College, University of Cambridge, CB3 9ET, UK
richard.bowman@cantab.net

10.1 Introduction

Chapters 1 to 4 explored the physical basis of optical tweezers, and Chapter 5 the many considerations involved in implementing such a system. In this chapter, we will explore some of the many uses to which optical tweezers have been put outside of microrheology (Chapters 6 to 9) and touch on the history of the technique's development. Starting with investigations of optical momentum, arguably the question that drove the development of optical tweezers, we will consider physical phenomena such as optical binding and hydrodynamic coupling. Biological experiments account for some of the most high-profile results to have been measured with optical tweezers, and this chapter will discuss a sample of these, notably work relating to molecular motors and synchronisation of swimming organisms. We will review progress in the use of optical tweezers as a scanning probe technique, including

Microrheology with Optical Tweezers: Principles and Applications
Edited by Manlio Tassieri

ISBN 978-981-4669-18-4 (Hardcover), 978-981-4669-19-1 (eBook)
www.panstanford.com

the use of optically actuated tools. Finally, developments in recent years have led to trapping and cooling of micro-objects in vacuum; this brings us rather neatly round to the science Arthur Ashkin was working towards when he developed the technique in the 1980s.

10.2 Optical Momentum

The origin of optical tweezers lies Arthur Ashkin's experiments in atom trapping. He used dielectric beads to explore the forces arising from light's momentum in preparation for exploiting the same effect to trap atoms (Ashkin, 1970; Ashkin et al., 1986). This original purpose, probing the momentum of light, is still an area of current experimentation and indeed theoretical debate. While Kepler first postulated that light might exert force on objects (as a result of observing comets), it was Abraham and Minkowsky who developed competing theories for the momentum of light as it passes through a dielectric (Leonhardt, 2006), a very relevant quantity for optical tweezers. Ashkin's earliest experiments (Ashkin, 1970) demonstrated that particles could be pushed by light, but quantitative agreement between the measured force and that expected based on the experimental geometry and the power of the laser has proved elusive until recently (Farré and Montes-Usategui, 2010). In these experiments, Farre et al. show that, from measuring the power and momentum distribution of light that has passed through a trapped particle, it is possible to recover the force acting on the particle, without any calibration of the spring constant of the optical trap. However, the precise relationship between the power of a light beam and its force has been the subject of debate (Leonhardt, 2006).

Momentum conservation would seem to give a simple, unambiguous answer for the force acting on a particle that deflects a light beam: the force is equal to the change in optical momentum per unit time. However, most optical trapping work is conducted in solution and so the light is propagating through a dielectric medium. Abraham's formulation suggests that the momentum per photon should be $\hbar k/n$ where k is the wavevector and n the refractive index of the medium, whereas Minkowsky suggests it is $n\hbar k$. Most of the time, the situation in optical tweezers means that the Minkowsky formulation should be the correct one: the lower momentum given by Abraham's formulation is what he ascribes to the light, and when conserving momentum we must also consider the momentum of the polarisation wave in the dielectric, which brings the two expressions into agreement. Some experiments have been performed with the aim of distinguishing between the Abraham and Minkowski forms of momentum but, in most situations commonly encountered in optical tweezers, the two theories give the same result.

Light carries linear momentum, but it can also have angular momentum. The most commonly known form of optical angular momentum is spin angular momentum, when the spins of the photons are all aligned in circularly polarised light. That this spin corresponds to a real angular momentum was elegantly demonstrated by He et al. in 1995 by optically trapping birefringent particles of calcite (He et al., 1995). When illuminated with circularly polarised light, these particles spin in the direction of the light's handedness, showing that it does indeed have a measurable momentum, leading to rotation as illustrated in Fig. 10.1. The light's polarisation state is changed after passing through the particle, and the particle's rotation is driven by the corresponding change

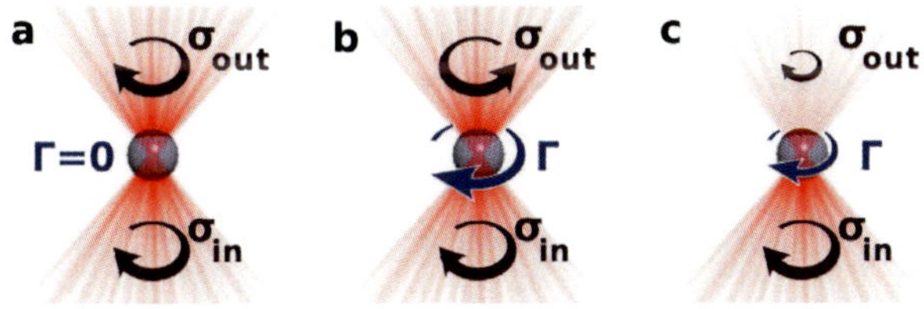

Figure 10.1 Spin angular momentum can rotate trapped particles. (a) A perfectly transparent, non-birefringent particle does not alter the light's spin angular momentum and so does not rotate. (b) A birefringent particle can change the light's spin and rotate very quickly (depending on the particle's geometry, it may or may not reverse the handedness, but as circularly polarised light represents the maximum spin angular momentum it should always rotate in the direction of the incident polarisation). (c) A partially absorbing particle picks up the angular momentum of the light it absorbs. Figure reproduced from (Bowman and Padgett, 2013) by kind permission of IoP Publishing.

in spin angular momentum. Further experiments showed that birefringence was not necessary for rotation, and in fact a weakly absorbing particle would also spin when trapped by a circularly polarised beam; in this case the polarisation state of the light is unchanged but, as some of the light has been absorbed, the particle still gains some angular momentum from the absorbed light (Simpson et al., 1997).

Spin angular momentum, however, is not the only form of angular momentum carried by light. A beam with appropriate spatial structure can carry orbital angular momentum. This is entirely independent of polarisation, and arises from the linear momentum of the light having an azimuthal component about some axis. Clearly any light beam could have orbital angular momentum about some axis but, for example in the case of Laguerre-Gaussian beams and higher-order Bessel beams, it is also possible to exhibit angular momentum about the axis of the light's propagation (Allen et al., 1992). This has been investigated in various geometries, leading to the rotation of objects about the centre of the beam. This rotation may coincide with rotation about the object's centre if the object is trapped in the centre of the beam or could result in an "orbit" of the trapped particles around the centre of the beam (O'Neil et al., 2002). The phase structure of a beam carrying orbital angular momentum generally has one or more singularities in the middle, necessitating a dark spot in the centre of the beam. The resultant ring-shaped structure means that particles can be trapped in a ring around the beam axis, which is where particles "orbit" in these experiments.

In order to demonstrate that spin and orbital angular momentum are separate quantities, it is possible to rotate particles with both simultaneously. Simpson and co-workers rotated a particle about its axis and added up or cancelled out the spin and orbital angular momentum contributions (Simpson et al., 1997), while O'Neil et al. showed that it is possible to have particles rotate about their axes due to spin momentum at the same time as orbiting the beam axis due to orbital angular momentum (O'Neil et al., 2002). Beads circulating in Laguerre-Gauss beams have even been used to probe the intensity structure of the beam; Curtis and Grier reported that the beam is not the size predicted by theory, as the most common method of producing these beams (using a forked

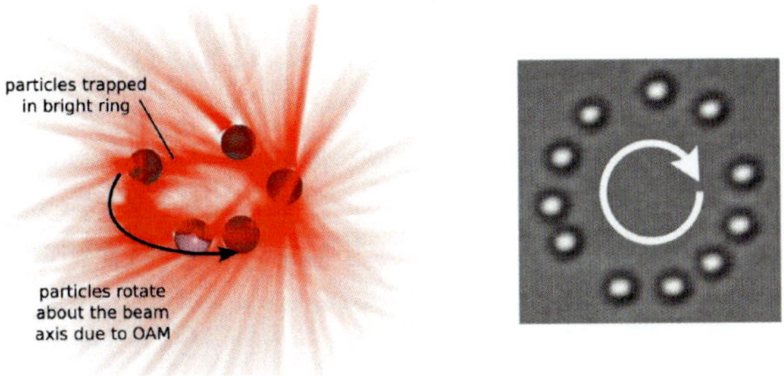

Figure 10.2 Particles can be trapped in the bright ring of a Laguerre-Gaussian beam, and the orbital angular momentum lost by light scattered from the beads causes them to circulate around the ring. Figure reproduced from (Bowman and Padgett, 2013) by kind permission of IoP Publishing.

hologram) approximates its intensity distribution in the Fourier plane with a Gaussian, neglecting the polynomial term that causes the centre of the beam to be dark (Curtis and Grier, 2003).

10.3 Statistical Mechanics

The circulation of particles in a Laguerre-Gaussian beam (illustrated in Fig. 10.2) is not only interesting because of what it can tell us about the orbital angular momentum of light, it also represents a very accessible model system for looking at statistical physics. Particles move in the optical force-field as a result of both the optical forces and thermal forces from the solvent molecules, and their consequent circulation can be thought of as a stochastic oscillator. The average period of the particle's circulation is well-defined by the magnitude of the azimuthal force and the viscous damping from the water, but thermal motion and the very low Reynolds number of the system means that there is a significant element of Brownian motion. The constant average force applied by the angular momentum of the beam leads, in essence, to a constant average drift velocity around the ring, on top of which Brownian motion is superimposed. This means that the phase of the oscillations is free to fluctuate even though their average period is known. If multiple beads are introduced to the system, their motion is not independent due to hydrodynamic interaction, and the beads in such a trap form an interesting, easily controlled chaotic system (Roichman et al., 2007). This phenomenon is not limited to Laguerre-Gauss beams, indeed it is possible to construct a wide range of traps that combine an extended trapping region with some average force (Roichman et al., 2008a).

Placing two such stochastic oscillators in close proximity has an interesting effect: due to hydrodynamic coupling, the oscillators will tend to synchronise with each other when their mean circulation rates are close (Di Leonardo et al., 2012). This has been postulated as a model for the synchronisation of bacterial flagella, and a number of other optical trap–based strategies have been employed to investigate synchronisation phenomena. Prior to the work with static, angular momentum-driven oscillators, Kotar et al. used a dynamic optical trapping system based on acousto-optic deflectors to set up two coupled stochastic oscillators and demonstrated synchronisation in their system, again designed to mimic the synchronisation of flagellae and cilia (Kotar et al., 2010).

Another statistical physics experiment to make use of dynamic optical traps is the experimental verification of the fluctuation theorems by Reid et al. (Carberry et al., 2004; Reid et al., 2004). This is a result that describes thermal fluctuations in non-equilibrium situations that can appear to violate the second law of thermodynamics on short timescales. These temporary decreases in entropy were observed and quantified by following the trajectory of optically trapped beads as the trap was switched repeatedly between two positions, and the expected exponential decrease of the magnitude of the fluctuations with time was observed. These fluctuations vanish on macroscopic time and length scales, which is why they can generally only be observed in microscopic systems.

Kramer's theorem is another important tenet of statistical mechanics, verified by McCann and colleagues using optical tweezers. The experiment simply analysed the motion of a trapped bead, held in a potential with two stable trapping points (McCann et al., 1999). This was formed by two laser foci, close enough that the particle could "hop" from one trap to the other through thermal fluctuations. Following from this, Curran et al. employed holographic optical tweezers to create two such dual-trap systems and showed that they hop in a correlated manner, much as two stochastic oscillators synchronise if placed close together (Curran et al., 2012).

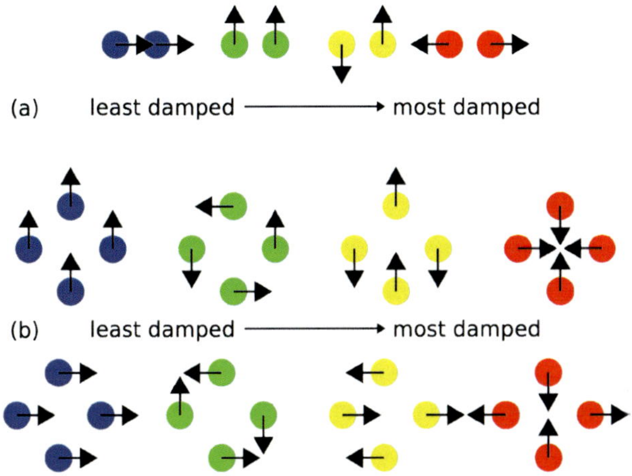

Figure 10.3 Eigenmodes, in two dimensions, of two (a) and four (b) beads in optical traps. They are arranged in approximate order of damping; common motion experiences the least hydrodynamic drag, while beads moving in opposite directions experience more friction than isolated beads.

Most of the above experiments depend on hydrodynamic coupling, first measured in optical tweezers by Meiners and co-workers (Meiners and Quake, 1999) and illustrated nicely by Polin et al. in 2006. They used a system comprising a linear array of optically trapped particles, in a water-glycerol mixture to slow the motion down sufficiently to track the particles with a video camera operating at 60 Hz (Polin et al., 2006). This showed viscous coupling between the beads, which has since been replicated at much higher speeds with modern cameras (Di Leonardo et al., 2007a) and interferometric detectors (Ruh et al., 2011). The motion of hydrodynamically coupled beads can be decomposed into eigenmodes, each with its own damping coefficient (Di Leonardo et al., 2007a). The simplest case of this, the motion of two beads in two dimensions, is shown in Fig. 10.3(a), but it can be extended to the motion of more particles and, indeed, more complicated objects (Simpson and Hanna, 2009). Common motion (i.e. beads moving in the same direction) results in a friction coefficient per bead that is lower than that for single beads moving, leading to the "pelotons" (groups of particles travelling together) seen in some of the experiments described in Section 10.3.

Damping coefficients for different eigenmodes can be measured in real time using high speed cameras (Carberry et al., 2010; Curran et al., 2010), and has been proposed as a microrheology technique that can distinguish effectively between proximity effects and viscosity changes, as some of the eigenmodes are affected more than others by the presence of a boundary near the trapped particles. An interesting biological manifestation of hydrodynamic coupling is the motion of *E. Coli* swimming near a flat surface: hydrodynamic coupling with the bacterium's hydrodynamic "image" in the surface (the flow field near a flat surface can be shown to be equivalent to the flow field given by two bacteria, where one is the reflection of the other in the surface) causes these bacteria to follow curved trajectories (Di Leonardo et al., 2011).

10.4 Optical Binding

Hydrodynamic forces are not the only long-range interactions that optically trapped particles can experience: if the trap geometry is carefully arranged, light scattered form one object can affect another. Sometimes, this causes the motion of multiple optically trapped particles to be highly correlated, with optical forces holding groups of particles in quite rigid configurations, much stiffer than the optical trap holding the group in one place (usually deliberately weak in these experiments). This effect, known as optical binding, was first investigated in 1989 by Burns and co-workers, using an extended optical trap (Burns et al., 1989, 1990). The trap was large and not stiff, so the particles were not optically trapped in three dimensions; instead, they were constrained against the microscope slide. Optical binding can be observed in dual-beam traps, and in this case the effects are often even stronger, as the beads are aligned along the direction of the scattered light (Tatarkova et al., 2002). Light entering a spherical particle can be brought to a sharp near-field focus just behind the particle, forming a strong trap for the next particle, and thus the particles form regularly-spaced chains as illustrated in Fig. 10.4. Using Bessel beams instead of Gaussian beams means that the focus is extended along the propagation axis, resulting in a trap with very low stiffness which is perfect for

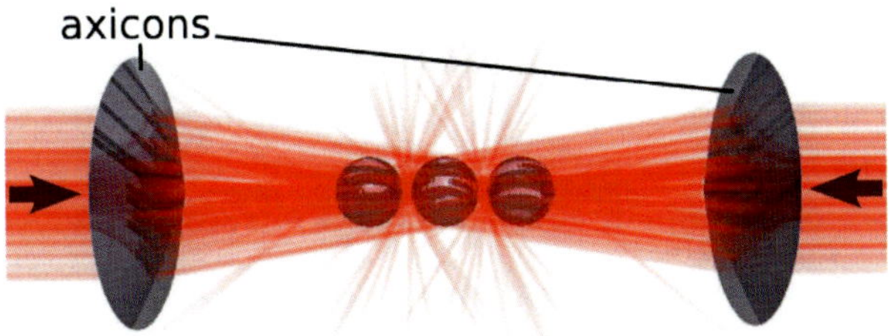

Figure 10.4 Three optically bound beads, guiding light from one bead to the next, in an extended optical trap formed by two counterpropagating Bessel beams. Figure reproduced from (Bowman and Padgett, 2013) by kind permission of IoP Publishing.

optical binding. Chains of many particles can be observed, and they exhibit surprising nonlinear dynamics which has been the subject of extensive simulation (Karásek and Zemánek, 2007; Taylor and Love, 2009).

10.5 Counterpropagating Traps

Counterpropagating-beam optical traps have been used in many optical binding experiments, but their utility goes far beyond those experiments. In fact, counterpropagating traps were first used by Ashkin in 1970 to cancel out the scattering force and allow a stable 3D optical trap to be formed from two low-NA beams (Ashkin, 1970). This trapping geometry has been used to produce long working distance traps, free from the constraints of high-NA microscope objectives, and has been combined with the Generalised Phase Contrast technique to create reconfigurable traps, analogous to holographic optical tweezers (Rodrigo et al., 2005a,b). The low-NA beams are compatible with lower magnification microscope objectives, allowing lower magnification objectives with larger fields of views. This permits larger-scale microassembly than with high-magnification objectives (Rodrigo et al., 2009), and imaging modes that employ multiple objectives (Perch-Nielsen et al., 2005).

Dual-beam traps can also be implemented using two optical fibres, as employed by Guck et al. on a microfluidic device for trapping single cells (Guck et al., 2001). The cell so trapped experiences a deformation force that "stretches" it along the axis of the optical trap (i.e. the direction of the two fibres), or equivalently a force that squeezes it towards the optical axis (stretching and squeezing are essentially equivalent to the Minkowski and Abraham interpretations). Due to the extremely stable geometry, such a trap can reach quite large forces, and by observing the deformation of cells under this stress it is possible to measure the elastic modulus of the cytoskeleton. This has been suggested as a potential diagnostic tool for cancer, an application which is aided by the ability to make such a trap on a low-cost microfluidic device, without an expensive high-NA microscope objective (Guck et al., 2005).

10.6 Single Molecule Studies

Perhaps the best known optical tweezers experiments since the technique's conception are those pertaining to motor proteins such as kinesins, myosins and dyneins. The first such experiment was carried out by Svoboda and Block using a specially-adapted optical trapping system that achieved unparalleled precision (Svoboda et al., 1993). By exploiting the polarisation optics used for differential interference contrast (DIC) imaging, they were able to achieve Angstrom resolution in position, and thus resolve the steps taken by a kinesin molecule as it pulled a bead along a microtubule. Their instrument, detailed in Fig. 10.5, used a Wollaston prism to create two very closely-spaced optical traps with orthogonal polarisations. These were then recombined by a second prism in the condenser assembly, such that the light from the two traps interfered, after passing through the trapped bead. By measuring the phase shift between the two foci, they could very accurately determine the particle's position along the shearing axis of the prisms and identify the molecule's discrete steps along the microtubule. The optical trap did more than just measure displacement: had the experiment simply watched beads being pulled along microtubules, Brownian motion would have blurred out the steps and only the average speed would have been measurable. By applying a force to the bead and making the molecular motor pull against the optical trap, the system

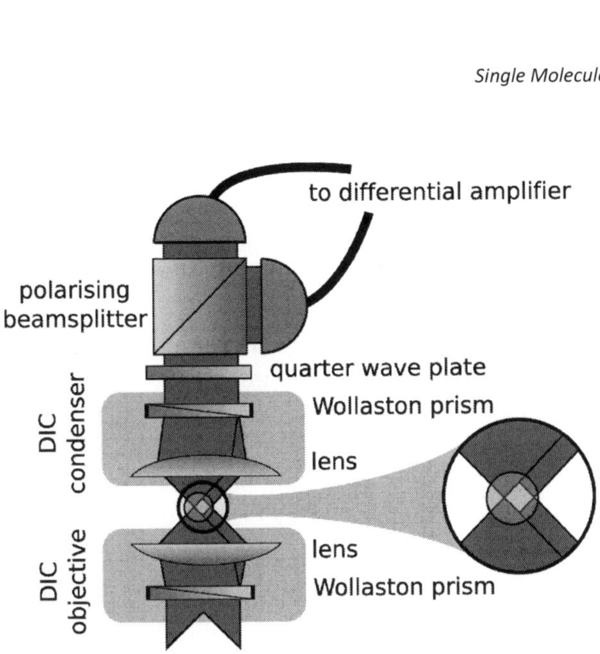

Figure 10.5 The optical trapping interferometer used by Svoboda and Block (Svoboda et al., 1993). Figure reproduced from (Bowman and Padgett, 2013) by kind permission of IoP Publishing.

was made stiffer and the steps were more easily resolved. Pulling against the protein also allowed them to measure its performance under different, controlled loads.

Kinesin is not the only motor protein to be investigated using optical tweezers, indeed all three major families of motor protein (kinesin, myosin and dynein) have been studied (Molloy et al., 1995; Sakakibara et al., 1999) as well as the rotary motor protein responsible for moving many bacterial flagellae (Pilizota et al., 2009). Besides motor proteins, optical tweezers have been used to investigate the properties of single strands of various biological macromolecules such as DNA (Bustamante et al., 2003) as well as the rheological properties of polymer solutions. By attaching beads to one or both ends of a strand, it can be stretched either between two optical traps or between one trap and a fixed surface (Smith et al., 1996; Wang et al., 1997). This allows force-extension curves to be plotted with exquisite resolution in force, orders of magnitude below what is generally achievable with atomic force microscopy (AFM) studies. DNA has been studied this way, and Smith et al. were able to successfully measure the force-extension curve for double stranded DNA. Perkins and co-workers used optical tweezers and fluorescence microscopy to directly visualise the reptation of a single strand of DNA, thus confirming at the single-molecule level many of the theories of polymer rheology we rely on today (Perkins et al., 1994a,b).

Using more traps and/or combining optical tweezers with fluorescence microscopy allows us to learn even more about the DNA under study, for example Noom et al. have shown it is possible to "knot" two strands of DNA together and, by sliding the knot along one strand of DNA, to identify bound proteins (Noom et al., 2007). Knots have also been tied in DNA by Bao and colleagues, who were interested in following the motion of the DNA as it relaxed out of the knotted state (Bao et al., 2003).

10.7 Scanning Probe Microscopy

A knot of DNA is not the only thing that can be used to "feel" out a sample, and indeed there have been a number of instances of optically trapped probes being used for scanning probe microscopy. The first instance of this technique was when Ghislain and co-workers used a shard of glass to scan over a surface and measure its relief (Ghislain and Webb, 1993). Unfortunately, poor resolution in tracking of the trapped shard and the unreliability of its fabrication meant that this technique was not to find extensive use, but subsequent realisations of the idea have proved more capable. Photonic force microscopy (PFM) is now an established technique that uses the Brownian motion of a small spherical probe to map out structures in 3D, though it is usually limited to transparent objects as otherwise the trapping beam and position detection are both distorted (Florin et al., 1997; Pralle et al., 1998, 1999; Rohrbach et al., 2004). Overcoming this limitation, Seitz et al. scanned 400 nm beads over various objects and, using aberration-corrected interferometric detection, were able to achieve resolutions limited by the bead's radius of curvature (Seitz et al., 2006). Aberration correction is particularly necessary when detecting the position of a trapped probe just above a sample, as the trapping light will be distorted by the sample. The small probe size also means that

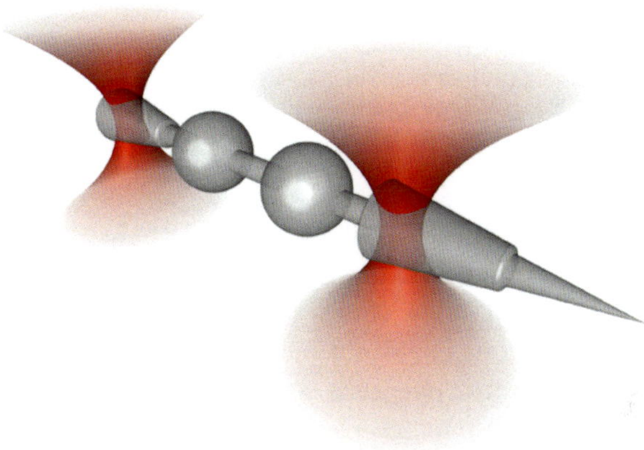

Figure 10.6 A microfabricated optical tool. Two-photon polymerisation was used to produce conical trapping handles, which function as constant-force optical springs. Spherical trapping handles can be used to manipulate the tool with higher stiffness, or to track its motion while it is trapped using the conical handles. Image courtesy of David Phillips.

there will necessarily be a significant amount of laser light passing through the sample even if it is beside, rather than underneath, the probe particle. Trapping light passing through the object under investigation could be a problem for absorbing or otherwise photosensitive samples, dubbed "opticution" by Ashkin when he first demonstrated the utility of optical tweezers for biological manipulations (Ashkin, 1997; Ashkin and Dziedzic, 1987).

Rodrigo and colleagues pursued an alternative solution to the problem of trapping close to a sample, by using optically trapped tools to spatially separate the traps from the object being probed (Rodrigo et al., 2005b). By constructing objects with a number of trapping "handles" and a sharp tip, it is possible to probe a sample mechanically without illuminating it with the trapping light. Phillips and co-workers were the first to use tools together with single-beam optical traps and position tracking to perform quantitative force and position measurement (Phillips et al., 2011, 2012b). Various different tool geometries and fabrication methods have been used, with the most versatile being formed by two-photon polymerisation. These tools can even have non-spherical trapping handles, which allow the stiffness of the optical traps to be tailored, for example to create a constant-force spring (Phillips et al., 2012b).

Two-photon polymerisation can create a wide variety of structures out of transparent polymer, and has in the past been used to fabricate waveguides of various different geometries. Palima et al. combined waveguides with optically actuated tools to give repositionable optical probes, able to guide light from a low-NA microscope objective to a sharp tip (Palima et al., 2012). This opens the possibility of probing nanoscale features that would usually require scanning near-field optical microscopy (SNOM), using a probe held in optical tweezers. Such a probe could be used in a closed sample cell to interrogate parts of a sample that are inaccessible to conventional SNOM, which requires a mechanical connection from an optical fibre all the way to the sample.

10.8 Vacuum Trapping and Cooling

Some recent works take the field of optical trapping full circle, returning to some of the fundamental physics that was Ashkin's original goal. While atom-trapping is now a well-established field, the trapping and cooling of dielectric microspheres in a vacuum is something that has only become possible in the last few years. Ashkin experimented with optically levitated particles in vacuum (Ashkin and Dziedzic, 1976), and did implement feedback stabilisation, though not fast enough to cool them to low temperatures (Ashkin and Dziedzic, 1977). Fast electronics now enable feedback control of optical traps with latencies below the velocity autocorrelation time, i.e. they can react with MHz bandwidth to damp the thermal motion of a trapped bead. As a bead in a vacuum is very well isolated from sources of heat, attenuating its motion cools the bead down to as low as 50 mK (Gieseler et al., 2012) or 1.5 mK (Li et al., 2011). This temperature corresponds to a thermal occupation number of about 3400. Further work using faster feedback control and lower-noise detection may be able to cool these mesoscopic particles considerably further. Once the particles have been cooled with the feedback system, the team plans to cool further by coupling the trapped sphere to an optical cavity, which should allow it to be cooled to its quantum ground state, making a very intriguing quantum system. Having a larger mass and more freedom of motion than any other quantum system, an optically trapped sphere in its quantum ground state could be regarded as a "Schrodinger's cat", a macroscopic system that is nonetheless in a superposition of states. It should be possible to test the relationship between gravity and quantum mechanics with this system, as its mass is large enough that gravitationally-induced decoherence should be observable on accessible timescales while the sphere is in free fall.

As well as being thermally isolated, beads in a vacuum experience very little friction. This has been exploited by Arita et al. to spin a birefringent particle faster than any other man-made spinning object (Arita et al., 2013). Trapping a vaterite sphere (vaterite is a spherical birefringent crystal) in a circularly polarised laser beam caused the particle to rotate at up to 10 MHz. It is not clear what should set the upper limit on speed in this system, the authors postulate the answer to this question will provide an insight into quantum frictional effects.

10.9 Conclusions

Optical tweezers have been used in a wide variety of contexts outside microrheology, across the physical and life sciences. Fields as diverse as statistical mechanics and molecular biology all have a role for the precise transduction of very small forces, torques and displacements; optical tweezers is the ideal tool for these measurements. Development of optical trapping technology continue to move the technique forward as a tool for both investigating fundamental physics and probing biological and chemical problems. As with microrheology, optical tweezers are maturing as a technology and there is plenty opportunity in the future for using these tools to learn more about the microscopic world.

Appendix

Evaluating the Fourier Transform*

R. Mike L. Evans

Department of Applied Mathematics, University of Leeds, Leeds LS2 9JT, UK

r.m.l.evans@leeds.ac.uk

A.1 Introduction

There are many different ways to do rheology—many different experimental techniques by which to measure the mechanical properties of viscoelastic materials. Some techniques measure a compliance, others measure a modulus. Some measurements are most conveniently expressed as a function of time, others as a function of frequency.

For instance, in a passive microrheology experiment, the time-dependent mean-square displacement of probe particles is measured as the particles perform Brownian motion within the viscoelastic medium. That increasing function of time is a measure of the fluid's time-dependent compliance (Mason et al., 1997; Waigh, 2005; Xu et al., 1998). The complance is also provided by a macroscopic creep experiment where a stress step is applied to a material, and its resulting strain is followed as time proceeds. A

*Much of the material in this appendix has appeared in (Evans, 2009).

Microrheology with Optical Tweezers: Principles and Applications
Edited by Manlio Tassieri

ISBN 978-981-4669-18-4 (Hardcover), 978-981-4669-19-1 (eBook)
www.panstanford.com

creep experiment is often the only feasible probe of the slowest relaxation modes of high-viscosity materials such as drilling muds or soft glassy materials including pastes and slurries.

The raw data generated by time-dependent compliance experiments appear quite different from the frequency-dependent storage and loss moduli ($G'(\omega)$ and $G''(\omega)$) measured in oscillatory rheometry. Nevertheless, so long as the measurements were performed in the linear regime, the two experiments in fact probe exactly the same physical properties of the material, so that the time-dependent compliance and the frequency-dependent modulus data contain the same information.

This can be easily understood if one imagines performing oscillatory measurements at two different frequencies simultaneously—a technique known as multi-wave rheology. Applying a high-frequency oscillatory stress superimposed on top of a low-frequency oscillatory stress causes the material to respond at both frequencies simultaneously, and the high- and low-frequency parts of its strain can be separately measured, giving its moduli at both frequencies. The experimenter could add in more frequencies, to measure more points on the spectrum at once. By adding up the appropriate amounts of all the odd harmonic frequencies, one can generate a square wave. A single stress step is just a square wave of infinitely low frequency, and is therefore equivalent to performing oscillatory measurements at *all* frequencies at once.

Since the data from creep and oscillatory rheometry contain equivalent information, it is equally acceptable to express their results by plotting a graph of compliance versus time, or of storage and loss moduli versus frequency. It is useful to express the results from both experiments in the same format, so that they can be directly compared. In practise, the frequency-dependent moduli have become the more standard rheological measure, so the need exists to convert creep compliance data into $G'(\omega)$ and $G''(\omega)$, and therein lies a problem. How do we convert the data accurately, i.e. without introducing artefacts?

In principle, the conversion is absolutely straightforward, since the frequency-dependent moduli are inversely proportional to the frequency-dependent compliance. So the time-dependent compliance must simply be Fourier-transformed. There are plenty of standard tools available for Fourier-analysing a signal, but the compliance signal from a step-stress experiment has several features that over-tax the approximations inherent in those tools, liberally decorating the output with artefacts.

A.2 Transforming from Time to Frequency with Minimal Artefacts

The main problem is one inherent in many experiments. We wish to process (in this case, to Fourier-transform) a signal that is a function of time, in this case the time-dependent compliance $J(t)$. It Fourier transform, the frequency-dependent compliance $\hat{J}(\omega)$, is defined by

$$\hat{J}(\omega) = \int_{-\infty}^{\infty} J(t)\, e^{-i\omega t} \mathrm{d}t \tag{A.1}$$

Unfortunately, we do not have access to the true function of time $J(t)$ since, in practice, the data have three shortcomings: the measurements begin some short but non-zero time after the stress step was applied (due to finite time resolution); the measurements are discrete, not continuous (also due to the temporal resolution); and the data terminate after some finite duration.

The mathematical recipe for a Fourier transform requires as input a function that is defined for *all* values of the time t. So we have a dilemma: what continuous function of t should we substitute into the mathematics instead of the experimental data? The set of possible choices is infinite, so care must be taken to avoid introducing subjective bias in judging the *least arbitrary* function consistent with the data. It would be unwise to evade the choice by entering the data into some proprietary spectrum analyser. The approximation scheme would then be hidden from the scientist, making it impossible to assess the extent of the artefacts.

Ones initial guess might be to replace the infinite-domain integration in Eq. A.1 by a finite summation over the data points $\{(t_k, J_k)\}$ thus:

$$J^{\text{guess}}(\omega) = \sum_{k=1}^{N} J_k\, e^{-i\omega t_k}.$$

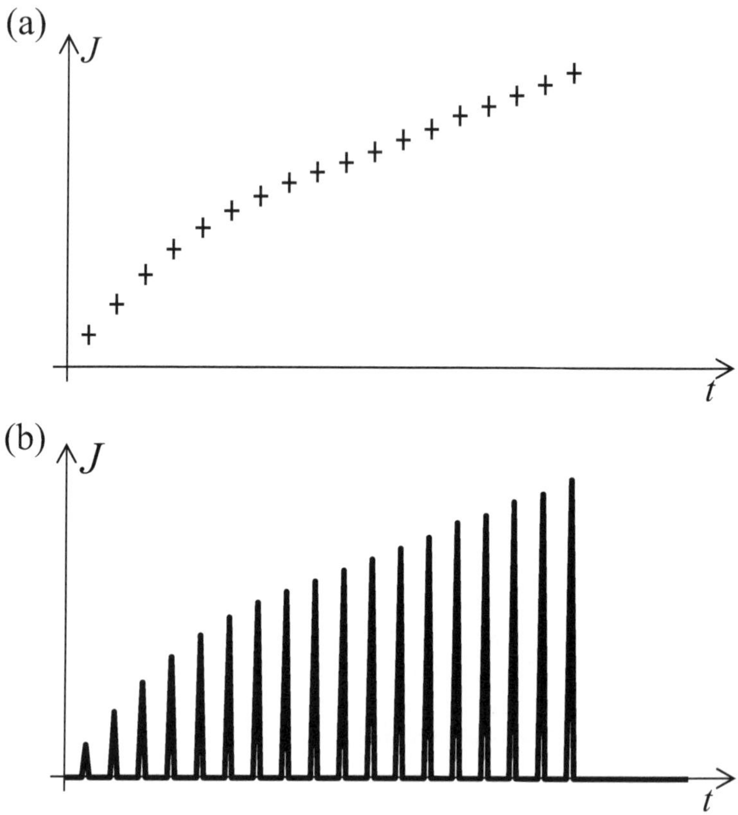

Figure A.1 (a) Sketch of an example data set from experimental measurements of a time-dependent compliance J. (b) Naïvely substituting the data directly into a Fourier integral (replacing the integral by a summation over the data points, $\sum_k J_k \exp(i\omega t_k)$) is tantamount to replacing the data by the function shown (a series of delta functions), and thus assuming that the compliance has a value of zero at all times except for those when a measurement was taken.

However, that choice would be equivalent to representing $J(t)$ by a function

$$J^{\text{guess}}(t) = \sum_{k=1}^{N} J_k \delta(t - t_k)$$

which is sum of Dirac delta functions, since substituting it into Eq. A.1 yields

$$\int_{-\infty}^{\infty} J^{\text{guess}}(t)\, e^{-i\omega t} \mathrm{d}t = \sum_{k=1}^{N} J_k \int_{-\infty}^{\infty} \delta(t - t_k)\, e^{-i\omega t} \mathrm{d}t = J^{\text{guess}}(\omega)$$

For a data-set such as that in Fig. A.1a, this choice would be represented by the function in Fig. A.1b. Since the true compliance is known to be a continuous, monotonically increasing function, there could hardly be a worst choice to represent it. Clearly, some interpolation and extrapolation is unavoidable.

Another common approach is to fit the data to a theoretical model of viscoelastic fluids, such as the generalised Maxwell model, in which a few Maxwell modes are added together to model relaxations in the fluid that occur at a range of different rates (Bird et al., 1987). Each mode is characterized by an amplitude and a time scale. Thus the experimental compliance data are used to establish the values of a set of fitting parameters—two per Maxwell mode—and the frequency-dependent moduli of that best-fit model can then be plotted.

Such a scheme blurs the distinction between theory and data. The resulting graphs of $G'(\omega)$ and $G''(\omega)$ can be legitimately interpreted as the output of a model, with the understanding that they therefore do not carry the weight of a true empirical observation.

Alternatively, without introducing any theoretical model, the time-dependent data could be fitted by any smooth fitting function, such as a *large* number of decaying exponentials (i.e. and inverse Laplace transform (Provencher, 1982)), or any other arbitrary choice. If this is done, the scientist must be aware that some arbitrariness is being introduced at that point.

Furthermore, a smooth function that does not pass through *every* data point would be a subjective corruption of the data. It is important to keep in mind the purpose of this exercise in interpolation and extrapolation, which is to allow us to Fourier-analyse the data, *not* to remove experimental noise. The data inevitably contain some noise, which should not be concealed, but should be faithfully reproduced, for the readers' scrutiny, in the resulting graphs of $G'(\omega)$ and $G''(\omega)$.

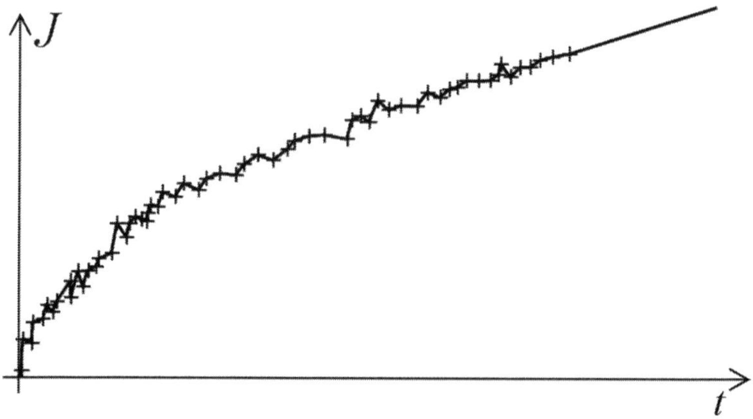

Figure A.2 A piecewise-linear interpolation between the data point, and also piece-wise linear extrapolations.

The simplest (and therefore arguably the least arbitrary) continuous function, passing through the data, would be a piecewise linear interpolation, as sketched in Fig. A.2.

To turn a finite data set into a function defined for all t, one must not only interpolate but also extrapolate to $t = 0$ and $t = \infty$, which can be done with two additional straight-line segments. If the compliance experiment was performed for sufficiently long to reach the terminal regime, where all elastic memory is lost, and the fluid flows with its long-time viscosity η, then the straight-line extrapolation to $t = \infty$ is uncontroversial. The final line-segment is given a gradient $1/\eta$, found by esimating the asymptote to the data by eye. Note that the human eye is the only good way to judge an asymptote to noisy or unpredictable data, since automated schemes are prone to error. Similarly, the value J_0 of the compliance at $t = 0$ must be carefully judged from a linear or log-linear plot. If, on the other hand, the data do not reach the terminal regime, then no amount of analysis can recover the frequency-dependent moduli from the incomplete experiment.

It is straightforward to write down the exact Fourier transform of the piecewise-linear fit, yielding a simple formula, found in (Evans et al., 2009) and reproduced here

$$G'(\omega) + iG''(\omega) = i\omega \Big/ \Bigg[i\omega J_0 + \left(1 - e^{-i\omega t_1}\right) \frac{(J_1 - J_0)}{t_1} + \frac{e^{-i\omega t_N}}{\eta} + \sum_{k=2}^{N} \left(\frac{J_k - J_{k-1}}{t_k - t_{k-1}} \right) \left(e^{-i\omega t_{k-1}} - e^{-i\omega t_k} \right) \Bigg] \quad \text{(A.2)}$$

in terms of the two parameters J_0 and η, and the N data points (t_k, J_k). Equation A.2 can be downloaded for various standard data processing packages from:

https://sites.google.com/site/manliotassieri/.

This formula will return an answer for any value of frequency ω that is substituted into it. But, of course, the data only contain information about a limited range of frequencies. Specifically, the frequency window depends on the resolution (of order t_1, the time of the first data point) and duration t_N of the data set, so that the formula returns real, scientifically valid information (including a faithful reproduction of the experimental noise) for frequencies in the range $t_N^{-1} < \omega < t_1^{-1}$. It can be usefully incorporated into real-time analysers, as discussed in (Yanagishima et al., 2011).

Outside that frequency range, it can return only artefact-dominated misinformation. Specifically, the function is then strongly influenced by the artificial features of the fit: the straight-line extrapolation to infinity and the sharp corners of the jagged function. It was found, in (Tassieri et al., 2012), that a substantial reduction in the size of the high-frequency artefacts, from which some high-frequency noise tends to spill over into the top of the experimental frequency range, can be achieved by an over-sampling technique. The technique involves first numerically interpolating between data points using a standard non-overshooting cubic spline, and then generating a new, oversampled data set, by sampling the interpolating function not only at the exact data points but also at a number of equally-spaced points in between. Equation A.2 is then applied to the oversampled data set.

The efficacy of Eq. A.2 has been demonstrated in a number of different experimental systems including a PAS melt (Evans, 2009), a polyisoprene melt in cone-and-plate geometry (Evans et al., 2009), magnetic microrheological measurements of F-actin solutions (a semi-felxible biopolymer) (Tassieri, 2007; Tassieri et al., 2012), and semidilute polyacrylamide solutions measured by both magnetic microrheology (Tassieri, 2007) and active microrheology with optical tweezers (Preece et al., 2011; Tassieri et al., 2010a). It has even been applied to non-equilibrium samples of aqueous wormlike micellar solutions (Willmer, 2010), since it does not require the sample to be at equilibrium. So long as the sample's moduli evolve only slowly—on a much longer time-scale that the measurement—the conversion remains valid. This makes creep the measurement of choice for slowly evolving non-equilibrium samples, as it can be performed more quickly than the equivalent oscillatory frequency sweep.

The formula should not be applied to non-linear rheology or to materials that are ageing rapidly on the timescale of the measurements. In both of those cases the dynamic moduli are anyway not uniquely defined, as their definitions must then take account of coupling between harmonics (Wilhelm et al., 1998) and of absolute time (Fielding et al., 2000).

References

Allan, D. W. (1966). Statistics of atomic frequency standards, *Proc. IEEE* **54**, 2, 221–230.

Allen, L., Beijersbergen, M. W., Spreeuw, R. J. C., and Woerdman, J. P. (1992). Orbital angular-momentum of light and the transformation of Laguerre-Gaussian laser modes, *Phys. Rev. A* **45**, 8185–8189.

Allen, L., Courtial, J., and Padgett, M. (1999). Matrix formulation for the propagation of light beams with orbital and spin angular momenta, *Phys. Rev. E* **60**, 7497–7503.

Apgar, J., Tseng, Y., Fedorov, E., Herwig, M., Almo, S., and Wirtz, D. (2000). Multiple-particle tracking measurements of heterogeneities in solutions of actin filaments and actin bundles, *Biophys. J.* **79**, 2, 1095–1106.

Arita, Y., Mazilu, M., and Dholakia, K. (2013). Laser-induced rotation and cooling of a trapped microgyroscope in vacuum, *Nat. Commun.* **4**, 2374.

Ashkin, A. (1970). Acceleration and trapping of particles by radiation pressure, *Phys. Rev. Lett.* **24**, 4, 156–159.

Ashkin, A. (1992). Force of a single-beam gradient laser trap on a dielectric sphere in the ray optics regine, *Biophys. J.* **61**, 2, 569–582.

Ashkin, A. (1997). Optical trapping and manipulation of neutral particles using lasers, *Proc. Natl. Acad. Sci. U. S. A.* **94**, 10, 4853–4860.

Ashkin, A., and Dziedzic, J. M. (1971). Optical levitation by radiation pressure, *Appl. Phys. Lett.* **19**, 8, 283–285.

Ashkin, A., and Dziedzic, J. M. (1976). Optical levitation in high vacuum, *Appl. Phys. Lett.* **28**, 6, 333–335.

Ashkin, A., and Dziedzic, J. M. (1977). Feedback stabilization of optically levitated particles, *Appl. Phys. Lett.* **30**, 4, 202–204.

Ashkin, A., and Dziedzic, J. M. (1987). Optical trapping and manipulation of viruses and bacteria, *Science* **235**, 4795, 1517–1520.

Ashkin, A., Dziedzic, J. M., Bjorkholm, J. E., and Chu, S. (1986). Observation of a single-beam gradient force optical trap for dielectric particles, *Opt. Lett.* **11**, 5, 288–290.

Ashkin, A., Dziedzic, J. M., and Yamane, T. (1987). Optical trapping and manipulation of single cells using infrared laser beams, *Nature* **330**, 769–771.

Ashkin, A., Schütze, K., Dziedzic, J. M., Euteneuer, U., and Schliwa, M. (1990). Force generation of organelle transport measured in vivo by an infrared laser trap, *Nature* **348**, 346–348.

Atakhorrami, M., Addas, K. M., and Schmidt, C. F. (2008). Twin optical traps for two-particle cross-correlation measurements: Eliminating cross-talk, *Rev. Sci. Instrum.* **79**, 043103.

Atakhorrami, M., Sulkowska, J. I., Addas, K. M., Koenderink, G. H., Tang, J. X., Levine, A. J., MacKintosh, F. C., and Schmidt, C. F. (2006). Correlated fluctuations of microparticles in viscoelastic solutions: Quantitative measurement of material properties by microrheology in the presence of optical traps, *Phys. Rev. E* **73**, 6, 061501.

Bao, X., Lee, H., and Quake, S. R. (2003). Behavior of complex knots in single dna molecules, *Phys. Rev. Lett.* **91**, 26, 265506.

Barnett, S. M., and Loudon, R. (2006). On the electromagnetic force on a dielectric medium, *J. Phys. B* **39**, 15, S671.

Barnett, S. M., and Zambrini, R. (2007). *Quantum Imaging* (Singapore: Springer), 284.

Beijersbergen, M. W., Allen, L., van der Veen, H., and Woerdman, J. P. (1993). Astigmatic laser mode converters and transfer of orbital angular momentum, *Opt. Commun.* **96**, 123–132.

Benito, D. C., Simpson, S. H., and Hanna, S. (2008). FDTD simulations of forces on particles during holographic assembly, *Opt. Express* **16**, 5, 2942–2957.

Berg-Sørensen, K., and Flyvbjerg, H. (2004). Power spectrum analysis for optical tweezers, *Rev. Sci. Instrum.* **75**, 3, 594–612.

Berne, B., and Pecora, R. (2000). *Dynamic Light Scattering: With Applications to Chemistry, Biology, and Physics* (New York: Wiley).

Berns, M. W., Aist, J. R., Wright, W. H., and Liang, H. (1992). Optical trapping in animal and fungal cells using a tunable, near-infrared titanium-sapphire laser, *Exp. Cell Res.* **198**, 375–378.

Bianchi, S., and Leonardo, R. D. (2010). Real-time optical micro-manipulation using optimized holograms generated on the GPU, *Comput. Phys. Commun.* **181**, 1444–1448.

Bingelyte, V., Leach, J., Courtial, J., and Padgett, M. J. (2003). Optically controlled three-dimensional rotation of microscopic objects, *Appl. Phys. Lett.* **82**, 829–831.

Bird, R., Curtiss, C., Armstrong, R., and Hassager, O. (1987). *Dynamics of Polymeric Liquids*, 2nd edn. (Wiley, New York).

Bishop, A. I., Nieminen, T. A., Heckenberg, N. R., and Rubinsztein-Dunlop, H. (2004). Optical microrheology using rotating laser-trapped particles, *Phys. Rev. Lett.* **92**, 19, 198104.

Block, S., Blair, D., and Berg, H. (1989). Compliance of bacterial flagella measured with optical tweezers, *Nature* **338**, 6215, 514–518.

Block, S. M. (1995). Nanometres and piconewtons: the macromolecular mechanics of kinesin, *Trends Cell. Biol.* **5**, 4, 169–175.

Bowman, R., D'Ambrosio, V., Rubino, E., Jedrkiewicz, O., Trapani, P. D., and Padgett, M. J. (2011). Optimisation of a low cost SLM for diffraction efficiency and ghost order suppression, *Eur. Phys. J. Special Topics* **199**, 149–158.

Bowman, R., Gibson, G., and Padgett, M. (2010). Particle tracking stereomicroscopy in optical tweezers: Control of trap shape, *Opt. Express* **18**, 11, 11785–11790.

Bowman, R. W., Gibson, G. M., Linnenberger, A., Phillips, D. B., Grieve, J. A., Carberry, D. M., Serati, S., Miles, M. J., and Padgett, M. J. (2014). "Red Tweezers": Fast, customisable hologram generation for optical tweezers, *Comput. Phys. Commun.* **185**, 1, 268–273.

Bowman, R. W., Gibson, G. M., Padgett, M. J., Saglimbeni, F., and Di Leonardo, R. (2013). Optical trapping at gigapascal pressures, *Phys. Rev. Lett.* **110**, 9, 095902.

Bowman, R. W., and Padgett, M. J. (2013). Optical trapping and binding, *Rep. Prog. Phys.* **76**, 2, 026401.

Brau, R., Ferrer, J., Lee, H., Castro, C., Tam, B., Tarsa, P., Matsudaira, P., Boyce, M., Kamm, R., and Lang, M. (2007). Passive and active microrheology with optical tweezers, *J. Opt. A-Pure Appl. Opt.* **9**, 8, SI, S103–S112.

Brouhard, G. J., Schek, H. T., and Hunt, A. J. (2003). Advanced optical tweezers for the study of cellular and molecular biomechanics, *IEEE Trans. Biomed. Eng.* **50**, 1, 121–125.

Burns, M. M., Fournier, J. M., and Golovchenko, J. A. (1989). Optical binding, *Phys. Rev. Lett.* **63**, 12, 1233–1236.

Burns, M. M., Fournier, J. M., and Golovchenko, J. A. (1990). Optical matter - crystallization and binding in intense optical-fields, *Science* **249**, 4970, 749–754.

Bustamante, C., Bryant, Z., and Smith, S. (2003). Ten years of tension: single-molecule dna mechanics, *Nature* **421**, 423–426.

Carberry, D. M., Reid, J., Wang, G. M., Sevick, E., Searles, D., and Evans, D. J. (2004). Fluctuations and irreversibility: An experimental demonstration of a second-law-like theorem using a colloidal particle held in an optical trap, *Phys. Rev. Lett.* **92**, 14, 140601.

Carberry, D. M., Simpson, S. H., Grieve, J. A., Wang, Y., Schafer, H., Steinhart, M., Bowman, R. W., Gibson, G., Padgett, M. J., Hanna, S., and Miles, M. J. (2010). Calibration of optically trapped nanotools, *Nanotechnology* **21**, 17, 175501.

Chaikin, P. M., and Lubensky, T. C. (2000). *Principles of Condensed Matter Physics*, Vol. 1 (Cambridge University Press, Cambridge).

Chen, D., Weeks, E., Crocker, J., Islam, M., Verma, R., Gruber, J., Levine, A., Lubensky, T., and Yodh, A. (2003). Rheological microscopy: Local mechanical properties from microrheology, *Phys. Rev. Lett.* **90**, 10, 108301.

Cicuta, P., and Donald, A. M. (2007). Microrheology: a review of the method and applications, *Soft Matter* **3**, 12, 1449–1455.

Corbitt, T., Ottaway, D., Innerhofer, E., Pelc, J., and Mavalvala, N. (2006). Measurement of radiation-pressure-induced optomechanical dynamics in a suspended fabry-perot cavity, *Phys. Rev. A* **74**, 2, 021802.

Coulomb, M. (1785). Premier mémoire sur l'électricité et le magnétisme, *Mémoires de Mathématique et de Physique tirés des registres de l'Académie Royale des Sciences*, 569–577.

Crocker, J. (1997). Measurement of the hydrodynamic corrections to the Brownian motion of two colloidal spheres, *J. Chem. Phys.* **106**, 7, 2837–2840.

Crocker, J., and Grier, D. (1994). Microscopic measurement of the pair interaction potential of charge-stabilized colloid, *Phys. Rev. Lett.* **73**, 2, 352–355.

Crocker, J., and Grier, D. (1996). Methods of digital video microscopy for colloidal studies, *J. Colloid Interface Sci.* **179**, 1, 298–310.

Crocker, J., Valentine, M., Weeks, E., Gisler, T., Kaplan, P., Yodh, A., and Weitz, D. (2000). Two-point microrheology of inhomogeneous soft materials, *Phys. Rev. Lett.* **85**, 4, 888–891.

Curran, A., Lee, M. P., Padgett, M. J., Cooper, J. M., and Di Leonardo, R. (2012). Partial synchronization of stochastic oscillators through hydrodynamic coupling, *Phys. Rev. Lett.* **108**, 24, 240601.

Curran, A., Tuohy, S., Aarts, D. G. A. L., Booth, M. J., Wilson, T., and Dullens, R. P. A. (2014). Decoupled and simultaneous three-dimensional imaging and optical manipulation through a single objective, *Optica* **1**, 4, 223–226.

Curran, A., Yao, A. M., Gibson, G., Bowman, R. W., Cooper, J. M., and Padgett, M. J. (2010). Real time characterization of hydrodynamics in optically trapped networks of micro-particles, *J. Biophotonics* **3**, 4, 244–251.

Curtis, J. E., and Grier, D. G. (2003). Structure of optical vortices, *Phys. Rev. Lett.* **90**, 13, 133901.

Curtis, J. E., Koss, B. A., and Grier, D. G. (2002). Dynamic holographic optical tweezers, *Opt. Commun.* **207**, 1-6, 169–175.

Curtis, J. E., Schmitz, C. H. J., and Spatz, J. P. (2005). Symmetry dependence of holograms for optical trapping, *Opt. Lett.* **30**, 16, 2086–2088.

Czerwinski, F., Richardson, A. C., and Oddershede, L. B. (2009). Quantifying noise in optical tweezers by Allan variance, *Opt. Express* **17**, 15, 13255–13269.

Dam, J. S., Perch-Nielsen, I. R., Palima, D., and Glückstad, J. (2008). Three-dimensional imaging in three-dimensional optical multi-beam micromanipulation, *Opt. Express* **16**, 10, 7244–7250.

Dam, J. S., Perch-Nielsen, I., Palima, D., and Glückstad, J. (2009). Multi-particle three-dimensional coordinate estimation in real-time optical manipulation, *J. Eur. Opt. Soc, Rapid Publ.* **4**, 09045.

Dasgupta, B., Tee, S., Crocker, J., Frisken, B., and Weitz, D. (2002). Microrheology of polyethylene oxide using diffusing wave spectroscopy and single scattering, *Phys. Rev. E* **65**, 5, 051505.

Davis, J. A., Cottrell, D. M., Campos, J., Yzuel, M. J., and Moreno, I. (1999). Encoding amplitude information onto phase-only filters, *Appl. Opt.* **38**, 23, 5004–5013.

Di Leonardo, R., Buzas, A., Kelemen, L., Vizsnyiczai, G., Oroszi, L., and Ormos, P. (2012). Hydrodynamic synchronization of light driven microrotors, *Phys. Rev. Lett.* **109**, 3, 034104.

Di Leonardo, R., Dell'Arciprete, D., Angelani, L., and Iebba, V. (2011). Swimming with an image, *Phys. Rev. Lett.* **106**, 3, 038101.

Di Leonardo, R., Ianni, F., and Ruocco, G. (2007). Computer generation of optimal holograms for optical trap arrays, *Opt. Express* **15**, 4, 1913–1922.

Di Leonardo, R., Keen, S., Leach, J., Saunter, C. D., Love, G. D., Ruocco, G., and Padgett, M. J. (2007a). Eigenmodes of a hydrodynamically coupled micron-size multiple-particle ring, *Phys. Rev. E* **76**, 6, 061402.

Di Leonardo, R., Leach, J., Mushfique, H., Cooper, J., Ruocco, G., and Padgett, M. (2006). Multipoint holographic optical velocimetry in microfluidic systems, *Phys. Rev. Lett.* **96**, 13, 134502.

Di Leonardo, R., Ruocco, G., Leach, J., Padgett, M. J., Wright, A. J., Girkin, J. M., Burnham, D. R., and McGloin, D. (2007b). Parametric resonance of optically trapped aerosols (vol. 99, art no. 010601, 2007), *Phys. Rev. Lett.* **99**, 2, 029902.

Dienerowitz, M., Gibson, G., Bowman, R., and Padgett, M. (2011). Holographic aberration correction: optimising the stiffness of an optical trap deep in the sample, *Opt. Express* **19**, 24, 24589–24595.

Dienerowitz, M., Gibson, G., Dienerowitz, F., and Padgett, M. (2012). Expanding the toolbox for nanoparticle trapping and spectroscopy with holographic optical tweezers, *J. Opt.* **14**, 045003.

Dienerowitz, M., Mazilu, M., and Dholakia, K. (2008). Optical manipulation of nanoparticles: a review, *J. Nanophotonics* **2**, 1, 021875–021875.

Dinsmore, A., Yodh, A., and Pine, D. (1996). Entropic control of particle motion using passive surface microstructures, *Nature* **383**, 6597, 239–242.

Dufresne, E., and Grier, D. (1998). Optical tweezer arrays and optical substrates created with diffractive optics, *Rev. Sci. Instrum.* **69**, 5, 1974–1977.

Dufresne, E., Spalding, G., Dearing, M., Sheets, S., and Grier, D. (2001). Computer-generated holographic optical tweezer arrays, *Rev. Sci. Instrum.* **72**, 3, 1810–1816.

Einstein, A. (1956). *Investigations on the Theory of the Brownian Movement* (Courier Dover Publications).

Evans, R., Tassieri, M., Auhl, D., and Waigh, T. A. (2009). Direct conversion of rheological compliance measurements into storage and loss moduli, *Phys. Rev. E* **80**, 1, 012501.

Evans, R. M. L. (2009). Transforming from time to frequency without artefacts, *Brit. Soc. Rheology Bull.* **50**, p. 76.

Farré, A., and Montes-Usategui, M. (2010). A force detection technique for single-beam optical traps based on direct measurement of light momentum changes, *Opt. Express* **18**, 11, 11955–11968.

Ferrari, E., Emiliani, V., Cojoc, D., Garbin, V., Zahid, M., Durieux, C., Coppey-Moisan, M., and Fabrizio, E. D. (2005). Biological samples micro-manipulation by means of optical tweezers, *Microelectron. Eng.* **78-79**, 575–581.

Ferry, J. D. (1980). *Viscoelastic Properties of Polymers*, 3rd edn. (Wiley), ISBN 0471048941.

Fielding, S., Sollich, P., and Cates, M. (2000). Aging and rheology in soft materials, *J. Rheol.* **44**, 2, 323–369.

Finer, J., Simmons, R., and Spudich, J. (1994). Single myosin molecule mechanics–piconewton forces and nanometer steps, *Nature* **368**, 6467, 113–119.

Florin, E.-L., Pralle, A., Hörber, J. K. H., and Stelzer, E. H. K. (1997). Photonic force microscope based on optical tweezers and two-photon excitation for biological applications, *J. Struct. Biol.* **119**, 2, 202–211.

Fournier, J. M. R., Burns, M. M., and Golovchenko, J. A. (1995). Writing difractive structures by optical trapping, in Benton, SA (ed.), *Practical Holography IX*, Vol. 2406, ISBN 0-8194-1753-X, 101–111, Practical Holography IX Conference, San Jose, CA.

Franosch, T., Grimm, M., Belushkin, M., Mor, F. M., Foffi, G., Forro, L., and Jeney, S. (2011). Resonances arising from hydrodynamic memory in brownian motion, *Nature* **478**, 7367, 85–88.

Friese, M., Nieminen, T., Heckenberg, N., and Rubinsztein-Dunlop, H. (1998). Optical alignment and spinning of laser-trapped microscopic particles, *Nature* **394**, 6691, 348–350.

Galajda, P., and Ormos, P. (2001). Complex micromachines produced and driven by light, *Appl. Phys. Lett.* **78**, 2, 249–251.

Gardiner, C. W. (2010). *Stochastic Methods: A Handbook for the Natural and Social Sciences* (Springer, Berlin).

Gerchberg, R. W., and Saxton, W. O. (1972). A practical algorithm for the determination of phase from image and diffraction plane pictures, *Optik* **35**, 2, 237–246.

Ghislain, L., and Webb, W. (1993). Scanning-force microscope based on an optical trap, *Opt. Lett.* **18**, 19, 1678–1680.

Gibson, G., Carberry, D. M., Whyte, G., Leach, J., Courtial, J., Jackson, J. C., Robert, D., Miles, M., and Padgett, M. (2008). Holographic assembly workstation for optical manipulation, *J. Opt. A: Pure Appl. Opt.* **10**, 044009.

Gibson, G. M., Bowman, R. W., Linnenberger, A., Dienerowitz, M., Phillips, D. B., Carberry, D. M., Miles, M. J., and Padgett, M. J. (2012). A compact holographic optical tweezers instrument, *Rev. Sci. Instrum.* **83**, 11, 113107.

Gibson, G. M., Leach, J., Keen, S., Wright, A. J., and Padgett, M. J. (2008). Measuring the accuracy of particle position and force in optical tweezers using high-speed video microscopy, *Opt. Express* **16**, 19, 14561–14570.

Gieseler, J., Deutsch, B., Quidant, R., and Novotny, L. (2012). Subkelvin parametric feedback cooling of a laser-trapped nanoparticle, *Phys. Rev. Lett.* **109**, 103603.

Gittes, F., and Schmidt, C. (1998). Interference model for back-focal-plane displacement detection in optical tweezers, *Opt. Lett.* **23**, 1, 7–9.

Gouy, L. G. (1890). Sur une propriete nouvelle des ondes lumineuses, *C. R. Acad. Sci. Paris* **110**, 1251.

Grier, D. (2003). A revolution in optical manipulation, *Nature* **424**, 6950, 810–816.

Griffiths, D. J. (1998). *Introduction to Electrodynamics*, 3rd edn. (Addison Wesley).

Guck, J., Ananthakrishnan, R., Cunningham, C., and Kas, J. (2002). Stretching biological cells with light, *J. Phys.-Condens. Mat.* **14**, 19, 4843–4856, 3rd International Workshop on Current Problems in Complex Fluids: Self-Assembling Systems, Oaxaca, Mexico, Jul 10–14, 2001.

Guck, J., Ananthakrishnan, R., Mahmood, H., Moon, T., Cunningham, C., and Kas, J. (2001). The optical stretcher: A novel laser tool to micromanipulate cells, *Biophys. J.* **81**, 2, 767–784.

Guck, J., Ananthakrishnan, R., Moon, T., Cunningham, C., and Kas, J. (2000). Optical deformability of soft biological dielectrics, *Phys. Rev. Lett.* **84**, 23, 5451–5454.

Guck, J., Schinkinger, S., Lincoln, B., Wottawah, F., Ebert, S., Romeyke, M., Lenz, D., Erickson, H., Ananthakrishnan, R., Mitchell, D., Kas, J., Ulvick, S., and Bilby, C. (2005). Optical deformability as an inherent cell marker for testing malignant transformation and metastatic competence, *Biophys. J.* **88**, 5, 3689–3698.

Hansen, P. M., Bhatia, V. K., Harrit, N., and Oddershede, L. (2005). Expanding the optical trapping range of gold nanoparticles, *Nano Lett.* **5**, 10, 1937–1942.

Happel, J., and Brenner, H. (1983). *Low Reynolds Number Hydrodynamics: With Special Applications to Particulate Media*, Mechanics of Fluids and Transport Processes (Springer Netherlands).

Harada, Y., and Asakura, T. (1996). Radiation forces on a dielectric sphere in the rayleigh scattering regime, *Opt. Commun.* **124**, 5, 529–541.

Hasan, S. W., Ghannam, M. T., and Esmail, N. (2010). Heavy crude oil viscosity reduction and rheology for pipeline transportation, *Fuel* **89**, 5, 1095–1100.

Hay, R. F., Gibson, G. M., Lee, M. P., Padgett, M. J., and Phillips, D. B. (2014). Four-directional stereo-microscopy for 3D particle tracking with real-time error evaluation, *Opt. Express* **22**, 15, 18662–18667.

Hayasaki, Y., Itoh, M., Yatagai, T., and Nishida, N. (1999). Nonmechanical optical manipulation of microparticle using spatial light modulator, *Opt. Rev.* **6**, 1, 24–27.

He, H., Friese, M. E. J., Heckenberg, N. R., and Rubinsztein-Dunlop, H. (1995). Direct observation of transfer of angular momentum to absorptive particles from a laser beam with a phase singularity, *Phys. Rev. Lett.* **75**, 5, 826–829.

Jackson, J. D. (1998). *Classical Electrodynamics*, 3rd edn. (John Wiley and Sons).

Jannasch, A., Mahamdeh, M., and Schaeffer, E. (2011). Inertial effects of a small brownian particle cause a colored power spectral density of thermal noise, *Phys. Rev. Lett.* **107**, 22, 228301.

Jordan, P., Leach, J., Padgett, M., Blackburn, P., Isaacs, N., Goksör, M., Hanstorp, D., Wright, A., Girkin, J., and Cooper, J. (2005). Creating permanent 3D arrangements of isolated cells using holographic optical tweezers, *Lab Chip* **5**, 1224–1228.

Jun, Y., Tripathy, S. K., Narayanareddy, B. R. J., Mattson-Hoss, M. K., and Gross, S. P. (2014). Calibration of Optical Tweezers for In Vivo Force Measurements: How do Different Approaches Compare? *Biophys. J.* **107**, 6, 1474–1484.

Karásek, V., and Zemánek, P. (2007). Analytical description of longitudinal optical binding of two spherical nanoparticles, *J. Opt. A* **9**, 8, S215.

Keen, S., Leach, J., Gibson, G., and Padgett, M. J. (2007). Comparison of a high-speed camera and a quadrant detector for measuring displacements in optical tweezers, *J. Opt. A-Pure Appl. Opt.* **9**, 8, S264–S266.

Keen, S., Yao, A., Leach, J., Leonardo, R. D., Saunter, C., Love, G., Cooper, J., and Padgett, M. (2009). Multipoint viscosity measurements in microfluidic channels using optical tweezers, *Lab Chip* **9**.

Klein, M., Andersson, M., Axner, O., and Fällman, E. (2007). Dual-trap technique for reduction of low-frequency noise in force measuring optical tweezers, *Appl. Opt.* **46**, 3, 405–412.

Kogelnik, H., and Li, T. (1966). Laser beams and resonators, *Appl. Opt.* **5**, 1550–1567.

Kohlrausch, R., and Weber, W. (1857). Avhandlingen ger könichtigen sächsischen, *Gesellschaft der Wissenschaft, Math.-Phys.* **3**, 219–292.

Kotar, J., Debono, L., Bruot, N., Box, S., Phillips, D., Simpson, S., Hanna, S., and Cicuta, P. (2013). Optimal hydrodynamic synchronization of colloidal rotors, *Phys. Rev. Lett.* **111**, 22, 228103.

Kotar, J., Leoni, M., Bassetti, B., Lagomarsino, M. C., and Cicuta, P. (2010). Hydrodynamic synchronization of colloidal oscillators, *Proc. Natl. Acad. Sci. U. S. A.* **107**, 17, 7669–7673.

Koumakis, N., and Di Leonardo, R. (2013). Stochastic hydrodynamic synchronization in rotating energy landscapes, *Phys. Rev. Lett.* **110**, 17, 174103.

Kramers, H. (1940). Brownian motion in a field of force and the diffusion model of chemical reactions, *Physica* **7**, 4, 284–304.

Kyoung, M., Karunwi, K., and Sheets, E. D. (2007). A versatile multimode microscope to probe and manipulate nanoparticles and biomolecules, *J. Microsc.* **225**, 137–146.

Lathi, B. (2004). *Linear Systems and Signals*, Oxford Series in Electrical and Computer Engineering (Oxford University Press, USA).

Leach, J., Howard, D., Roberts, S., Gibson, G., Gothard, D., Cooper, J., Shakesheff, K., Padgett, M., and Buttery, L. (2009). Manipulation of live mouse embryonic stem cells using holographic optical tweezers, *J. Mod. Opt.* **56**, 4.

Leach, J., Mushfique, H., di Leonardo, R., Padgett, M., and Cooper, J. (2006). An optically driven pump for microfluidics, *Lab Chip* **6**, 6, 735–739.

Leach, J., Wulff, K., Sinclair, G., Jordan, P., Courtial, J., Thomson, L., Gibson, G., Karunwi, K., Cooper, J., Laczik, Z. J., and Padgett, M. (2006). Interactive approach to optical tweezers control, *Appl. Opt.* **45**, 897–903.

Lee, M. P., Gibson, G. M., Bowman, R., Bernet, S., Ritsch-Marte, M., Phillips, D. B., and Padgett, M. J. (2013). A multi-modal stereo microscope based on a spatial light modulator, *Opt. Express* **21**, 14, 16541–16551.

Lee, M. P., Gibson, G. M., Phillips, D., Padgett, M. J., and Tassieri, M. (2014). Dynamic stereo microscopy for studying particle sedimentation, *Opt. Express* **22**, 4, 4671–4677, doi:10.1364/OE.22.004671.

Lee, M. P., and Padgett, M. J. (2012). Optical tweezers: a light touch, *J. Microsc.* **248**, 3, 219–222.

Lemons, D. S., and Gythiel, A. (1997). Paul langevin's 1908 paper "on the theory of brownian motion" ["sur la théorie du mouvement brownien," c. r. acad. sci. (paris) 146, 530–533 (1908)], *Am. J. Phys.* **65**, 11, 1079–1081.

Leonhardt, U. (2006). Optics: Momentum in an uncertain light, *Nature* **444**, 7121, 823–824.

Levine, A., and Lubensky, T. (2000). One- and two-particle microrheology, *Phys. Rev. Lett.* **85**, 8, 1774–1777.

Li, T., Kheifets, S., and Raizen, M. G. (2011). MilliKelvin cooling of an optically trapped microsphere in vacuum, *Nat. Phys.* **7**, 7, 527–530.

Liesener, J., Reicherter, M., Haist, T., and Tiziani, H. (2000). Multi-functional optical tweezers using computer-generated holograms, *Opt. Commun.* **185**, 1-3, 77–82.

Lincoln, B., Schinkinger, S., Travis, K., Wottawah, F., Ebert, S., Sauer, F., and Guck, J. (2007a). Reconfigurable microfluidic integration of a dual-beam laser trap with biomedical applications, *Biomed. Microdevices* **9**, 5, 703–710.

Lincoln, B., Wottawah, F., Schinkinger, S., Ebert, S., and Guck, J. (2007b). High-throughput rheological measurements with an optical stretcher, in Wang, YL and Discher, DE (ed.), *Cell Mechanics, Methods in Cell Biology*, Vol. 83 (Elsevier Inc.), 397–423.

Liu, Y., Cheng, D. K., Sonek, G. J., Berns, M. W., Chapman, C. F., and Tromberg, B. J. (1995). Evidence for localized cell heating induced by infrared optical tweezers, *Biophys. J.* **68**, 2137–2144.

M-Badosa, E., M-Usategui, M., Carnicer, A., Andilla, J., Pleguezuelos, E., and Juvells, I. (2007). Design strategies for optimizing holographic optical tweezers set-ups, *J. Opt. A: Pure Appl. Opt.* **9**, S267–S277.

MacDonald, M., Spalding, G., and Dholakia, K. (2003). Microfluidic sorting in an optical lattice, *Nature* **426**, 6965, 421–424.

Mason, T. (2000). Estimating the viscoelastic moduli of complex fluids using the generalized Stokes-Einstein equation, *Rheol. Acta* **39**, 4, 371–378.

Mason, T., Ganesan, K., vanZanten, J., Wirtz, D., and Kuo, S. (1997). Particle tracking microrheology of complex fluids, *Phys. Rev. Lett.* **79**, 17, 3282–3285.

Mason, T., and Weitz, D. (1995). Optical measurements of frequency-dependent linear viscoelastic moduli of complex fluids, *Phys. Rev. Lett.* **74**, 7, 1250–1253.

Maxwell, J. C. (1855). VIII: On faraday's lines of force, *Transactions of the Cambridge Philosophical Society* **x**, I.

Maxwell, J. C. (1862). On physical lines of force. Pt. III, *Scientific Papers* **1**, 498.

Maxwell, J. C. (1864). A dynamical theory of the electromagnetic field, *Roy. Soc. Trans.* **CLV**, 526.

Maxwell, J. C. (1873). *A Treatise on Electricity and Magnetism*, Vol. 2 (Clarendon Press).

Mazolli, A., Neto, P. M., and Nussenzveig, H. (2003). Theory of trapping forces in optical tweezers, *Proc. R. Soc. A* **459**, 2040, 3021–3041.

McCann, L., Dykman, M., and Golding, B. (1999). Thermally activated transitions in a bistable three-dimensional optical trap, *Nature* **402**, 6763, 785–787.

Mehta, A., Rief, M., Spudich, J., Smith, D., and Simmons, R. (1999). Single-molecule biomechanics with optical methods, *Science* **283**, 5408, 1689–1695.

Meiners, J., and Quake, S. (1999). Direct measurement of hydrodynamic cross correlations between two particles in an external potential, *Phys. Rev. Lett.* **82**, 10, 2211–2214.

Meiners, J.-C., and Quake, S. R. (2000). Femtonewton force spectroscopy of single extended DNA molecules, *Phys. Rev. Lett.* **84**, 21, 5014–5017.

Meyer, A., Marshall, A., Bush, B., and Furst, E. (2006). Laser tweezer microrheology of a colloidal suspension, *J. Rheol.* **50**, 1, 77–92.

Milne, G., Rhodes, D., MacDonald, M., and Dholakia, K. (2007). Fractionation of polydisperse colloid with acousto-optically generated potential energy landscapes, *Opt. Lett.* **32**, 9, 1144–1146.

Moffitt, J. R., Chemla, Y. R., Izhaky, D., and Bustamante, C. (2006). Differential detection of dual traps improves the spatial resolution of optical tweezers, *Proc. Natl. Acad. Sci. U. S. A.* **103**, 24, 9006–9011.

Moffitt, J. R., Chemla, Y. R., Smith, S. B., and Bustamante, C. (2008). Recent advances in optical tweezers, *Annu. Rev. Biochem.* **77**, 205–228.

Molloy, J. E., Burns, J. E., Kendrick-Jones, J., Tregear, R. T., and White, D. C. S. (1995). Movement and force produced by a single myosin head, *Nature* **378**, 6553, 209–212.

Molloy, J. E., and Padgett, M. J. (2002). Lights, action: optical tweezers, *Contemp. Phys.* **43**, 4, 241–258.

Morse, D. (1998). Viscoelasticity of concentrated isotropic solutions of semiflexible polymers. 2. Linear response, *Macromolecules* **31**, 20, 7044–7067.

Mushfique, H., Leach, J., Di Leonardo, R., Padgett, M. J., and Cooper, J. M. (2008a). Optically driven pumps and flow sensors for microfluidic systems, *Proceedings of the Institution of Mechanical Engineers Part C-Journal of Mechanical Engineering Science* **222**, 5, 829–837.

Mushfique, H., Leach, J., Yin, H., Di Leonardo, R., Padgett, M. J., and Cooper, J. M. (2008b). 3D mapping of microfluidic flow in laboratory-on-a-chip structures using optical tweezers, *Anal. Chem.* **80**, 11, 4237–4240.

Neto, P. M., and Nussenzveig, H. (2000). Theory of optical tweezers, *Europhys. Lett.* **50**, 5, 702.

Neuman, K., and Block, S. (2004). Optical trapping, *Rev. Sci. Instrum.* **75**, 9, 2787–2809.

Nieminen, T. A., Loke, V. L., Stilgoe, A. B., Knöner, G., Brańczyk, A. M., Heckenberg, N. R., and Rubinsztein-Dunlop, H. (2007). Optical tweezers computational toolbox, *J. Opt. A* **9**, 8, S196.

Nijenhuis, N., Mizuno, D., Spaan, J. A. E., and Schmidt, C. F. (2009). Viscoelastic response of a model endothelial glycocalyx, *Phys. Biol.* **6**, 2, 025014.

Noom, M., Van Den Broek, B., Van Mameren, J., and Wuite, G. (2007). Visualizing single dna-bound proteins using dna as a scanning probe, *Nat. Methods* **4**, 12, 1031–1036.

Novotny, L., Bian, R., and Xie, X. (1997). Theory of nanometric optical tweezers, *Phys. Rev. Lett.* **79**, 4, 645–648.

O'Neil, A. T., MacVicar, I., Allen, L., and Padgett, M. J. (2002). Intrinsic and extrinsic nature of the orbital angular momentum of a light beam, *Phys. Rev. Lett.* **88**, 5, 053601.

Ota, T., Sugiura, T., Kawata, S., Booth, M. J., Neil, M. A. A., Juškaitis, R., and Wilson, T. (2003). Enhancement of laser trapping force by spherical aberration correction using a deformable mirror, *Jpn. J. Appl. Phys.* **42**, L701–L703.

Otto, O., Czerwinski, F., Gornall, J. L., Stober, G., Oddershede, L. B., Seidel, R., and Keyser, U. F. (2010). Real-time particle tracking at 10,000 fps using optical fiber illumination, *Opt. Express* **18**, 22, 22722–22733.

Otto, O., Gutsche, C., Kremer, F., and Keyser, U. F. (2008). Optical tweezers with 2.5kHz bandwidth video detection for single-colloid electrophoresis, *Rev. Sci. Instrum.* **79**, 023710.

Padgett, M., and Bowman, R. (2011). Tweezers with a twist, *Nature Photon.* **5**, 6, 343–348.

Palima, D., Bañas, A. R., Vizsnyiczai, G., Kelemen, L., Ormos, P., and Glückstad, J. (2012). Wave-guided optical waveguides, *Opt. Express* **20**, 3, 2004–2014.

Palmer, A., Mason, T., Xu, J., Kuo, S., and Wirtz, D. (1999). Diffusing wave spectroscopy microrheology of actin filament networks, *Biophys. J.* **76**, 2, 1063–1071.

Papagiannopoulos, A. (2005). The structure and dynamics of polyelectrolyte combs, *University of Leeds Ph.D. Thesis.*

Papagiannopoulos, A., Fernyhough, C., and Waigh, T. (2005a). The microrheology of polystyrene sulfonate combs in aqueous solution, *J. Chem. Phys.* **123**, 21, 214904.

Papagiannopoulos, A., Waigh, T., Fluerasu, A., Fernyhough, C., and Madsen, A. (2005b). Microrheology of polymeric solutions using x-ray photon correlation spectroscopy, *J. Phys.-Condens. Mat.* **17**, 25, L279–L285.

Papagiannopoulos, A., Waigh, T. A., Hardingham, T., and Heinrich, M. (2006). Solution structure and dynamics of cartilage aggrecan, *Biomacromolecules* **7**, 7, 2162–2172.

Papagiannopoulos, A., Waigh, T. A., and Hardingham, T. E. (2008). The viscoelasticity of self-assembled proteoglycan combs, *Faraday Discuss.* **139**, 337–357.

Parkin, S. J., Knöner, G., Nieminen, T. A., Heckenberg, N. R., and Rubinsztein-Dunlop, H. (2007). Picoliter viscometry using optically rotated particles, *Phys. Rev. E* **76**, 4, 041507.

Paterson, L., MacDonald, M., Arlt, J., Sibbett, W., Bryant, P., and Dholakia, K. (2001). Controlled rotation of optically trapped microscopic particles, *Science* **292**, 5518, 912–914.

Perch-Nielsen, I. R., Rodrigo, P., and Glückstad, J. (2005). Real-time interactive 3D manipulation of particles viewed in two orthogonal observation planes, *Opt. Express* **13**, 8, 2852–2857.

Perkins, T. T., Quake, S. R., Smith, D. E., and Chu, S. (1994a). Relaxation of a single DNA molecule observed by optical microscopy, *Science* **264**, 5160, 822–826.

Perkins, T. T., Smith, D. E., and Chu, S. (1994b). Direct observation of tube-like motion of a single polymer-chain, *Science* **264**, 5160, 819–822.

Pesce, G., De Luca, A. C., Rusciano, G., Netti, P. A., Fusco, S., and Sasso, A. (2009). Microrheology of complex fluids using optical tweezers: a comparison with macrorheological measurements, *J. Opt. A-Pure Appl. Opt.* **11**, 3, 034016.

Pesce, G., Volpe, G., Chiara De Luca, A., Rusciano, G., and Volpe, G. (2009). Quantitative assessment of non-conservative radiation forces in an optical trap, *EPL* **86**, 3, 38002.

Peterman, E. J. G., Gittes, F., and Schmidt, C. F. (2003). Laser-induced heating in optical traps, *Biophys. J.* **84**, 2.

Phillips, D., Grieve, J., Olof, S., Kocher, S., Bowman, R., Padgett, M., Miles, M., and Carberry, D. (2011). Surface imaging using holographic optical tweezers, *Nanotechnology* **22**, 28, 285503.

Phillips, D., Padgett, M., Hanna, S., Ho, Y.-L., Carberry, D., Miles, M., and Simpson, S. (2014). Shape-induced force fields in optical trapping, *Nature Photon.* **8**, 400–405.

Phillips, D. B., Gibson, G. M., Bowman, R., Padgett, M. J., Hanna, S., Carberry, D. M., Miles, M. J., and Simpson, S. H. (2012a). An optically actuated surface scanning probe, *Opt. Express* **20**, 28, 29679–29693.

Phillips, D. B., Simpson, S., Grieve, J., Bowman, R., Gibson, G., Padgett, M., Rarity, J., Hanna, S., Miles, M., and Carberry, D. (2012b). Force sensing with a shaped dielectric micro-tool, *Europhys. Lett.* **99**, 5, 58004.

Pilizota, T., Brown, M. T., Leake, M. C., Branch, R. W., Berry, R. M., and Armitage, J. P. (2009). A molecular brake, not a clutch, stops the rhodobacter sphaeroides flagellar motor, *Proc. Natl. Acad. Sci. U. S. A.* **106**, 28, 11582–11587.

Pine, D., Weitz, D., Chaikin, P., and Herbolzheimer, E. (1988). Diffusing-wave spectroscopy, *Phys. Rev. Lett.* **60**, 12, 1134–1137.

Polin, M., Grier, D. G., and Quake, S. R. (2006). Anomalous vibrational dispersion in holographically trapped colloidal arrays, *Physical Review Letters* **96**, 8, 088101.

Pralle, A., Florin, E.-L., Stelzer, E. H. K., and Hörber, J. K. H. (1998). Local viscosity probed by photonic force microscopy, *Appl. Phys. A* **66**, 71–73.

Pralle, A., Prummer, M., Florin, E.-L., Stelzer, E. H. K., and Hörber, J. K. H. (1999). Three-dimensional high-resolution particle tracking for optical tweezers by forward scattered light, *Microsc. Res. Techniq.* **44**, 5, 378–386.

Preece, D., Bowman, R., Linnenberger, A., Gibson, G., Serati, S., and Padgett, M. (2009). Increasing trap stiffness with position clamping in holographic optical tweezers, *Opt. Express* **17**, 25, 22718–22725.

Preece, D., Warren, R., Evans, R., Gibson, G. M., Padgett, M. J., Cooper, J. M., and Tassieri, M. (2011). Optical tweezers: wideband microrheology, *J. Opt.* **13**, 4, 044022.

Provencher, S. (1982). Contin: A general purpose constrained regularization program for contin: A general purpose constrained regularization program for inverting noisy linear algebraic and integral equations, *Comput. Phys. Commun.* **27**, 3, 229–242.

Reicherter, M., Haist, T., Wagemann, E., and Tiziani, H. (1999). Optical particle trapping with computer-generated holograms written on a liquid-crystal display, *Opt. Lett.* **24**, 9, 608–610.

Reicherter, M., Zwick, S., Haist, T., Kohler, C., Tiziani, H., and Osten, W. (2006). Fast digital hologram generation and adaptive force measurement in liquid-crystal-display-based holographic tweezers, *Appl. Opt.* **45**, 5, 888–896.

Reid, J., Carberry, D. M., Wang, G. M., Sevick, E., Evans, D. J., and Searles, D. (2004). Reversibility in nonequilibrium trajectories of an optically trapped particle, *Phys. Rev. E* **70**, 1, 016111.

Reihani, S. N. S., and Oddershede, L. B. (2007). Optimizing immersion media refractive index improves optical trapping by compensating spherical aberrations, *Opt. Lett.* **32**, 14, 1998–2000.

Risken, H. (1996). *The Fokker-Planck Equation: Methods of Solution and Applications* (Springer, Berlin).

Rodrigo, P., Daria, V. R., and Glückstad, J. (2005a). Four-dimensional optical manipulation of colloidal particles, *Appl. Phys. Lett.* **86**, 7, 074103.

Rodrigo, P., Gammelgaard, L., Bøggild, P., Perch-Nielsen, I., and Glückstad, J. (2005b). Actuation of microfabricated tools using multiple GPC-based counterpropagating-beam traps, *Opt. Express* **13**, 18, 6899–6904.

Rodrigo, P., Kelemen, L., Palima, D., Alonzo, C. A., Ormos, P., and Glückstad, J. (2009). Optical microassembly platform for constructing reconfigurable microenvironments for biomedical studies, *Opt. Express* **17**, 8, 6578–6583.

Rohrbach, A., and Stelzer, E. H. K. (2002). Three-dimensional position detection of optically trapped dielectric particles, *J. Appl. Phys.* **91**, 8, 5474–5488.

Rohrbach, A., Tischer, C., Neumayer, D., Florin, E.-L., and Stelzer, E. H. K. (2004). Trapping and tracking a local probe with a photonic force microscope, *Rev. Sci. Instrum.* **75**, 6, 2197–2210.

Roichman, Y., Grier, D. G., and Zaslavsky, G. (2007). Anomalous collective dynamics in optically driven colloidal rings, *Phys. Rev. E* **75**, 2, 020401.

Roichman, Y., Sun, B., Roichman, Y., Amato-Grill, J., and Grier, D. G. (2008a). Optical forces arising from phase gradients, *Phys. Rev. Lett.* **100**, 1, 013602.

Roichman, Y., Sun, B., Stolarski, A., and Grier, D. G. (2008b). Influence of nonconservative optical forces on the dynamics of optically trapped colloidal spheres: the fountain of probability, *Phy. Rev. Lett.* **101**, 12, 128301.

Rømer, O. (1676). Demonstration touchant le mouvement de la lumière trouvé par m. römer de l'académie royal des science, *Le Journal des Sçavans (Dec. 1676)*, 276.

Ruh, D., Traenkle, B., and Rohrbach, A. (2011). Fast parallel interferometric 3D tracking of numerous optically trapped particles and their hydrodynamic interaction, *Opt. Express* **19**, 22, 21627–21642.

Sakakibara, H., Kojima, H., Sakai, Y., Katayama, E., and Oiwa, K. (1999). Inner-arm dynein c of chlamydomonas flagella is a single-headed processive motor, *Nature* **400**, 6744, 586–590.

Sasaki, K., Koshioka, M., Misawa, H., Kitamura, N., and Masuhara, H. (1991). Pattern formation and flow control of fine particles by laser-scanning micromanipulation, *Opt. Lett.* **16**, 19, 1463–1465.

Sasaki, K., Koshioka, M., Misawa, H., Kitamura, N., and Masuhara, H. (1992). Optical trapping of a metal particle and a water droplet by a scanning laser beam, *Appl. Phys. Lett.* **60**, 7, 807–809.

Savin, T., and Doyle, P. (2005). Static and dynamic errors in particle tracking microrheology, *Biophys. J.* **88**, 1, 623–638.

Schmitz, C. H. J., Spatz, J. P., and Curtis, J. E. (2005). High-precision steering of multiple holographic optical traps, *Opt. Express* **13**, 21, 8678–8685.

Schnurr, B., Gittes, F., MacKintosh, F., and Schmidt, C. (1997). Determining microscopic viscoelasticity in flexible and semiflexible polymer networks from thermal fluctuations, *Macromolecules* **30**, 25, 7781–7792.

Schönle, A., and Hell, S. W. (1998). Heating by absorption in the focus of an objective lens, *Opt. Lett.* **23**, 5, 325–327.

Seitz, P. C., Stelzer, E. H. K., and Rohrbach, A. (2006). Interferometric tracking of optically trapped probes behind structured surfaces: a phase correction method, *Appl. Opt.* **45**, 28, 7309–7315.

Selhuber-Unkel, C., Zins, I., Schubert, O., Sönnichsen, C., and Oddershede, L. B. (2008). Quantitative optical trapping of single gold nanorods, *Nano Lett.* **8**, 9, 2998–3003.

Shannon, C. (1949). Communication in the presence of noise, *Proc. IRE* **37**, 1, 10–21.

Sharma, R. C., Papagiannopoulos, A., and Waigh, T. A. (2008). Optical coherence tomography picorheology of biopolymer solutions, *Appl. Phys. Lett.* **92**, 17, doi:10.1063/1.2917578.

Sheard, B. S., Gray, M. B., Mow-Lowry, C. M., McClelland, D. E., and Whitcomb, S. E. (2004). Observation and characterization of an optical spring, *Phys. Rev. A* **69**, 5, 051801.

Siegman, A. E. (1990). *Lasers* (University Science Books, U.S.).

Simmons, R. M., Finer, J. T., Chu, S., and Spudich, J. A. (1996). Quantitative measurements of force and displacement using an optical trap, *Biophys. J.* **70**, 1813–1822.

Simpson, N., Dholakia, K., Allen, L., and Padgett, M. J. (1997). Mechanical equivalence of spin and orbital angular momentum of light: An optical spanner, *Opt. Lett.* **22**, 1, 52–54.

Simpson, S. (2014). Inhomogeneous and anisotropic particles in optical traps: Physical behaviour and applications, *J. Quant. Spectrosc. Radiat. Trans.* **146**, 81–99.

Simpson, S. H., Benito, D. C., and Hanna, S. (2007). Polarization-induced torque in optical traps, *Phys. Rev. A* **76**, 4, 043408.

Simpson, S. H., and Hanna, S. (2007). Optical trapping of spheroidal particles in Gaussian beams, *JOSA A* **24**, 2, 430–443.

Simpson, S. H., and Hanna, S. (2009). Thermal motion of a holographically trapped SPM-like probe, *Nanotechnology* **20**, 39, 395710.

Simpson, S. H., and Hanna, S. (2010). Holographic optical trapping of microrods and nanowires, *JOSA A* **27**, 6, 1255–1264.

Sinclair, G., Jordan, P., Courtial, J., Padgett, M., Cooper, J., and Laczik, Z. J. (2004a). Assembly of 3-dimensional structures using programmable holographic optical tweezers, *Opt. Express* **12**, 5475–5480.

Sinclair, G., Jordan, P., Leach, J., Padgett, M. J., and Cooper, J. (2004b). Defining the trapping limits of holographical optical tweezers, *J. Mod. Opt.* **51**, 3, 409–414.

Sinclair, G., Leach, J., Jordan, P., Gibson, G., Yao, E., Laczik, Z. J., Padgett, M. J., and Courtial, J. (2004c). Interactive application in holographic optical tweezers of a multi-plane Gerchberg-Saxton algorithm for three-dimensional light shaping, *Opt. Express* **12**, 8, 1665–1670.

Smith, R. L., Spalding, G. C., Dholakia, K., and MacDonald, M. P. (2007). Colloidal sorting in dynamic optical lattices, *J. Opt. A-Pure Appl. Opt.* **9**, 8, SI, S134–S138.

Smith, S., Cui, Y., and Bustamante, C. (1996). Overstretching B-DNA: The elastic response of individual double-stranded and single-stranded dna molecules, *Science* **271**, 5250, 795–799.

Spruijt, E. (2010). Relaxation dynamics at different time scales in electrostatic complexes: Time-salt superposition, *Phys. Rev. Lett.* **105**, 20, 208301.

Squires, T. M., and Mason, T. G. (2010). Fluid Mechanics of Microrheology, *Annu. Rev. Fluid Mech.* **42**, 413–438.

Starrs, L., and Bartlett, P. (2003). One- and two-point micro-rheology of viscoelastic media, *J. Phys.-Condens. Mat.* **15**, 1, S251–S256.

Stokes, J. R., and Davies, G. A. (2007). Viscoelasticity of human whole saliva collected after acid and mechanical stimulation, *Biorheology* **44**, 3, 141–160.

Svoboda, K., and Block, S. (1994a). Biological applications of optical forces, *Annu. Rev. Biophys. Biomol. Struct.* **23**, 247–285.

Svoboda, K., and Block, S. M. (1994b). Optical trapping of metallic rayleigh particles, *Opt. Lett.* **19**, 13, 930–932.

Svoboda, K., Schmidt, C. F., Schnapp, B., and Block, S. M. (1993). Direct observation of kinesin stepping by optical trapping interferometry, *Nature* **365**, 6448, 721–727.

Swartzlander Jr, G. A., Peterson, T. J., Artusio-Glimpse, A. B., and Raisanen, A. D. (2011). Stable optical lift, *Nature Photon.* **5**, 1, 48–51.

Tassieri, M. (2007). Microrheology of semiflexible polymers, *University of Leeds Ph.D. Thesis*.

Tassieri, M., Gibson, G. M., Evans, R., Yao, A. M., Warren, R., Padgett, M. J., and Cooper, J. M. (2010a). Measuring storage and loss moduli using optical tweezers: Broadband microrheology, *Phys. Rev. E* **81**, 2, 026308.

Tassieri, M., Waigh, T., Trinick, J., Aggeli, A., and Evans, R. (2010b). Analysis of the linear viscoelasticity of polyelectrolytes by magnetic microrheometry-Pulsed creep experiments and the one particle response, *J. Rheol.* **54**, 1, 117–131.

Tassieri, M., Evans, R., Warren, R. L., Bailey, N. J., and Cooper, J. M. (2012). Microrheology with optical tweezers: data analysis, *New J. Phys.* **14**, 11, 115032.

Tatarkova, S., Carruthers, A., and Dholakia, K. (2002). One-dimensional optically bound arrays of microscopic particles, *Phys. Rev. Lett.* **89**, 28, 283901.

Taylor, J. M., and Love, G. D. (2009). Optical binding mechanisms: a conceptual model for Gaussian beam traps, *Opt. Express* **17**, 17, 15381–15389.

Theofanidou, E., Wilson, L., Hossack, W. J., and Arlt, J. (2004). Spherical aberration correction for optical tweezers, *Opt. Commun.* **236**, 145–150.

Tolic-Norrelykke, S. F., Schaeffer, E., Howard, J., Pavone, F. S., Juelicher, F., and Flyvbjerg, H. (2006). Calibration of optical tweezers with positional detection in the back focal plane, *Rev. Sci. Instrum.* **77**, 10, 103101.

Tskhovrebova, L., Trinick, J., Sleep, J., and Simmons, R. (1997). Elasticity and unfolding of single molecules of the giant muscle protein titin, *Nature* **387**, 6630, 308–312.

Tsuda, Y., Mori, O., Funase, R., Sawada, H., Yamamoto, T., Saiki, T., Endo, T., and Kawaguchi, J. (2011). Flight status of IKAROS deep space solar sail demonstrator, *Acta Astronaut.* **69**, 9, 833–840.

Uhlenbeck, G. E., and Ornstein, L. S. (1930). On the theory of the brownian motion, *Phys. Rev.* **36**, 5, 823.

Van Kampen, N. G. (2007). *Stochastic Processes in Physics and Chemistry* (North-Holland, Amsterdam).

Visscher, K., Brakenhoff, G. J., and Krol, J. J. (1993). Micromanipulation by "multiple" optical traps created by a single fast scanning trap integrated with the bilateral confocal scanning laser microscope, *Cytometry* **14**, 105–114.

Visscher, K., Gross, S. P., and Block, S. M. (1996). Construction of multiple-beam optical traps with nanometer-resolution position sensing, *IEEE J. Sel. Topics Quantum Electron.* **2**, 4, 1066–1076.

Waigh, T. (2005). Microrheology of complex fluids, *Rep. Prog. Phys.* **68**, 3, 685–742.

Wallin, A. E., Ojala, H., Hæggström, E., and Tuma, R. (2008). Stiffer optical tweezers through real-time feedback control, *Appl. Phys. Lett.* **92**, 224104.

Wang, F., Toe, W. J., Lee, W. M., McGloin, D., Gao, Q., Tan, H. H., Jagadish, C., and Reece, P. J. (2013). Resolving stable axial trapping points of nanowires in an optical tweezers using photoluminescence mapping, *Nano Lett.* **13**, 3, 1185–1191.

Wang, M., Yin, H., Landick, R., Gelles, J., and Block, S. (1997). Stretching DNA with optical tweezers, *Biophys. J.* **72**, 3, 1335–1346.

Weitz, D., and Pine, D. (1993). *Dynamic Light Scattering* (Oxford University Press, USA).

Wilhelm, M., Maring, D., and Spiess, H. (1998). Fourier-transform rheology, *Rheol. Acta* **37**, 4, 399–405.

Willmer, D. (2010). Non-equilibrium polymeric complex fluids, *Nottingham Trent University Ph.D. Thesis.*

Wirtz, D. (2009). Particle-tracking microrheology of living cells: principles and applications, *Annu. Rev. Biophys.* **38**, 301–326.

Wu, P., Huang, R., Tischer, C., Jonas, A., and Florin, E.-L. (2009). Direct Measurement of the Nonconservative Force Field Generated by Optical Tweezers, *Phys. Rev. Lett.* **103**, 10, 108101.

Wulff, K. D., Cole, D. G., Clark, R. L., Leonardo, R. D., Leach, J., Cooper, J., Gibson, G., and Padgett, M. J. (2006). Aberration correction in holographic optical tweezers, *Opt. Express* **14**, 4169–4174.

Xu, J., Viasnoff, V., and Wirtz, D. (1998). Compliance of actin filament networks measured by particle-tracking microrheology and diffusing wave spectroscopy, *Rheol. Acta* **37**, 4, 387–398.

Yanagishima, T., Frenkel, D., Kotar, J., and Eiser, E. (2011). Real-time monitoring of complex moduli from micro-rheology, *J. Physi.-Condens. Mat.* **23**, 19, 194118.

Yao, A., Tassieri, M., Padgett, M., and Cooper, J. (2009a). Microrheology with optical tweezers, *Lab Chip* **9**, 17, 2568–2575.

Yao, A. M., Keen, S. A. J., Burnham, D. R., Leach, J., Di Leonardo, R., McGloin, D., and Padgett, M. J. (2009b). Underdamped modes in a hydrodynamically coupled microparticle system, *New J. Phys.* **11**, 053007.

Yoon, Y.-Z., Kotar, J., Yoon, G., and Cicuta, P. (2008). The nonlinear mechanical response of the red blood cell, *Phys. Biol.* **5**, 3, 036007.

Zemánek, P., Jonáš, A., Šrámek, L., and Liška, M. (1999). Optical trapping of nanoparticles and microparticles by a Gaussian standing wave, *Opt. Lett.* **24**, 21, 1448–1450.

Index